I0568420

BEYOND BRAIN ROT

How to Stop Scrolling Your Life Away

AXEL BURLIN

Published by Glimmer Books, Albany, NY
Book website: beyondbrainrot.com
For author or publisher inquiries, contact: axel@glimmerbooks.com

ISBN 979-8-218-83331-2
First Edition Paperback

Contents

Introduction

I scrolled years of my life away. Every night I hunched over screens in my dark room, bingeing on social media, videos, forums, news feeds, and fantasy games. I could not stop myself. It felt like I was trapped in a pinball machine, endlessly ricocheting between snippets of addictive content.

During the day, my mind floated in a distracted haze, while my body shambled through the motions like a zombie, doing the bare minimum needed to return to my devices. There was no room for creativity, spontaneity, or peace—only consumption. *Just click, click, tap, tap, and smudge, smudge the shiny screens.*

Yet deep down, I knew something was wrong. Something within me wanted to *do stuff*: to adventure in the real world, to read leather-bound books, to wander in a garden—to escape the brain rot that dominated my 21st century life.

I remember standing in the shower one night (the only place I couldn't bring my screens) after frittering away yet another day online. As the steaming water rolled down my back and I stared at the ceramic wall tiles with crusty eyes, my blood started to tingle. I got angry. Was I really this weak-willed and pitiful? Doomed to waste my youth online? To lead a life half-lived of suffocating comfort, wasted potential, and mid-conversation phone-glances?

Far from being some minor first-world problem, the stakes of my internet addiction were life itself. The only thing I truly owned was my limited time on this planet. *This* present moment—this steamy ceramic-wall-tile-staring—was *it*. Life is a stretch of conscious experience, then it's over. So when online platforms hijacked this conscious experience, they hijacked my only life.

In that moment, I decided I would beat my internet addiction or die trying. I would free myself no matter what it took.

I just didn't realize it would take me four years, and many more angry showers, to do so.

At first I thought I just needed a physical product. I bought a kSafe—a plastic box for locking up your phone—and an off-brand version with a small hole that lets you answer any emergency phone calls. The problem was that I never really put my phone in the boxes, always coming up with excuses (*just one more YouTube video*, or, *what if my friend texts me?*). After a few days, they sat unused in the corner of my room.

Next, I bought a cheap flip phone (a "dumb phone"). It took just one day to realize how painfully impractical it was. Smartphones do have plenty of helpful functions, after all, with apps like maps, Uber, and email—they're not *all* bad. The flip phone also turned my texts green, ruining iMessage group chats and making me the laughingstock of my friends. And I only had a tiny plastic keyboard to defend myself. Most importantly, I simply pulled out my laptop and binged there instead.

But I never gave up. I tried everything, no matter how unconventional. I once even crafted a pair of anti-screen gloves—thin gloves that killed the conductivity of my fingertips, so touch screens wouldn't respond. I added a locking mechanism on the wrists, preventing me from easily slipping them off. Once finished, I remember looking at my gloved hands and thinking: *Really? This is nothing more than a 21st century chastity belt. How humiliating.* So I scrapped that idea as well.

After exhausting hardware-based solutions, I pivoted to software. Silicon Valley had gotten us into this mess; it was time to use their own weapons against them. I bought site-blocking software to keep me from distracting sites... But within a couple days I found a disable button and circumvented the blocks in two clicks. All software has these escape hatches: no third-party app can completely commandeer your devices—and in a moment of weakness, the tech-savvy addict will surely find them.

I tried countless screen time apps, including one that grew a digital tree the less I used my phone. Suffice it to say my tree withered and died very quickly. I once even partnered with a coder friend to build an app that spammed the user with angry notifications if they scrolled for too long. It was horrendously obnoxious, a Frankenstein's monster that even I, as its creator, could not tolerate. So I scrapped that as well.

Finally (and desperately), I turned to the dusty, analog world of books. I read every major title on the topic of internet addiction and its underlying neuroscience. Interestingly, this was the only method that actually helped. The books recognized that the solution to tech overload was not even more tech. They encouraged readers to slow down and reevaluate their digital habits on a more philosophical level—to address the cause, not just the symptoms. Through some of these books, I was able to quit for weeks at a time and felt gloriously alive.

Yet of all the books I read, not one allowed me to escape my internet addiction for good. For several reasons: first, all were written by authors aged 40 years and older. They didn't grow up addicted to their devices and didn't understand all the platforms, or why we *really* use them.

Second, none of them offered practical solutions beyond a handful of tips and tricks gleaned from tech-savvy nephews: use black-and-white mode, delete some apps, buy a watch. Instead, the books were stuffed with animal studies, Mark Zuckerberg's life story, and meandering anecdotes about Henry David Thoreau.

They then inevitably concluded, "the tech giants are too powerful, they need to be regulated. In the meantime, I too struggle with my digital devices, so you're not alone." Which is nice and all, but really, is the only way we can reclaim our lives through the ballot box? The average age of Congress is 60, and it's full of pockets lined by tech lobbyists, so I wouldn't count on it.

None of these books led me to escape my habits. But they gave me the material, and the material percolated and recombined in my head for a very long time before I discovered a new method of escape. I still remember it: I was sitting at home with pen and paper, sketching out a series of assumptions which, when taken to their logical conclusion, were a knockout punch to my addictive urges. Like a math problem with an elegant final solution. It all became clear to me: the *why*, the *what*, and the *how* that no one else was talking about. Most of all I felt the tragedy of the countless precious hours of my youth flushed down the skibidi toilet.* All of it had been an illusion, a fun house of mirrors that exploited my brain.

Once I put my new theory into practice, it actually worked. My screen time plummeted and my sense of autonomy returned, without a single gimmick or product involved. And the solution was unquestionably permanent.

Finally, I had found the escape from the prison of internet addiction that I'd been seeking for so long. It was so clean and simple that it felt like finding a key to the cell door right in my pocket. I didn't need to steal floor plans, bang at the prison bars, or scrape at the cell floor with a rusty spoon. I could just calmly walk to the door, slide my key into the lock, and walk out. Now I was in charge, not my devices: the internet served me and not the other way around.

---

* A YouTube Shorts series with billions of views (and middle school teacher complaints about students screaming "skibidi" in class).

Once I had discovered this method, I also knew I needed to share it with the world, with anyone who wants to free themselves from the same addiction.

Part One of this book defines the problem, because the first step in any battle, as Sun Tzu famously said, is to "know the enemy." Many other methods do not define exactly what the problem is that you need to remove from your life, and therefore give you no way to measure success. The internet obviously has lots of helpful uses, so you need to understand what to cut out and what to keep.

Part Two then shows you how to escape through a transformative new method that restructures your mindset. There is no tedious, step-by-step list of instructions or gimmicky settings to tweak on your phone—just a fundamental rewriting of the faulty beliefs that lead to addictive behaviors. You may feel the urge to skip to this section, but Part One is necessary to build a foundation and know exactly what you're quitting. This book can save thousands of hours of your life and works best when read through sequentially.

Throughout the book, I often reference studies and make factual claims; all sources are listed in the Notes section at the end to preserve readability.

You are about to embark on an exciting transformation of your everyday life. Soon, you too will free yourself from internet addiction.

PART I

The Problem

The Rise of Brain Rot

The internet is everywhere and has many practical uses, so what exactly is the problem? As in war, we must know who our enemy is before we can fight them. Otherwise, it's like wearing a blindfold in a sword fight and swinging wildly around, vainly hoping to hit the target.

So who is our enemy?

The superficial answer is brain rot: hyper-stimulating content that we regret consuming afterwards. Brain rot offers no practical benefit. As a form of low-brow entertainment, its main goal is to stimulate us emotionally. It appeals to our monkey brains with drama, violence, sexual or weird situations that we find irresistible.

Our higher selves know this type of content is not good for us. Think about it: if we spent our screen time studying, enjoying beautiful art, writing love letters, starting a business, or otherwise advancing our goals, we'd be happy. But that's not how we spend it. In practice, we spend our screen time on clickbait, political scandals, random comment sections, and a distant friend-of-a-friend's thirst traps—in other words, brain rot.

Yet brain rot itself is not the enemy. As media theorist Marshall McLuhan, best known for coining the phrase "the medium is the message," wrote in 1964: "the 'content' of a medium is like the juicy piece of meat carried by the burglar to distract the watchdog of the mind." It's the medium through which we consume the brain rot that matters. To see the problem, the true enemy we face, we must look beyond brain rot.

You see, brain rot has always existed. People have consumed extreme, salacious, low-brow entertainment for millennia; it's only human. However, the technological medium through which brain rot is transmitted has changed drastically over time. And that has made all the difference.

In the past, brain rot was dormant: a small, flickering flame confined to the edges of culture. Then, over the course of just a few centuries, tech and media profiteers found this flame, this appetite that has always been in our psyches, and fanned it. Advances in technology allowed them to feed us ever-increasing doses of brain rot, ever-more efficiently. They fed the flame kindling and wood until it became a powerful, irresistible bonfire that sucked all the oxygen out of the room.

So let's take a walk through history to examine the evolution of brain rot's delivery method. After all, in Cicero's words, "To be ignorant of what occurred before you were born is to remain always a child." To truly understand our enemy, we must understand its origin story.

Stage 1—Live Performance
from the dawn of humanity until the 15th century

For almost all of human history, entertainment was exclusively performed live. It was a fleeting delicacy constrained by the physicality of the medium. It meant sitting around the campfire, listening to an elder tell tales or the resident bone-flutist play a tune. Almost no one knew how to read or write (only a few monks and cloistered academics), so it was an oral culture of memorized poetry, performance, and storytelling.

Much of this storytelling, as far as we can tell, involved tales of heroes (Gilgamesh, Odysseus, Beowulf, etc.), genealogies, moral parables, religious stories and creation myths. But of course, that elder at the campfire might also have some brain rot up their tunic-sleeves. Consider Greek mythology, for example, with Aphrodite's many melodramatic affairs and Dionysus' wine and orgy-filled bacchanals. But if no one had good storytelling skills (or memory), you were out of luck.

If you lived in Ancient Rome in the era of *panem et circenses* ("bread and circuses") you might get to see gladiator battles, chariot races, and simulated hunts. In the medieval era, perhaps troupes of traveling actors, jousting tournaments, and bards singing in the tavern. Or the classic brain-rot-fun of throwing rotten vegetables at pilloried criminals. Regardless, we relied entirely on live performances by the people in our physical vicinity. There was no entertainment-on-demand; it was all confined to real people and the limitations of time and space.

Stage 2–The Written Word
from the 15th to the 20th century

This changed in 1440 when Johannes Gutenberg invented the printing press and modern mass media entertainment was born. Text could now be replicated quickly and cheaply. It would take a couple centuries to unleash its full entertainment value (the first mass-produced novel, *Don Quixote,* was published in 1605), but once it was, people could keep books at home and entertain themselves. And they proved voracious consumers.

Bibliographers estimate that 13 million books were printed in the 15th century, 217 million in the 16th, 532 million in the 17th, and nearly one billion in the 18th. The proliferation of the printed word gave the public literary greats like Charles Dickens, Jane Austen, and the Brontë sisters. Critics of the written word were rare, and books were here to stay.

Yet the advance in printing also provided a new medium for brain rot to exploit. This happened most notably in the 1830s with the invention of the penny press. Benjamin Day realized that if he sold ads in his newspaper, *The New York Sun,* he could lower the price of the newspaper itself. With a wide enough readership, advertisers would pay him money, and he could keep lowering the price until the newspaper was practically free—only a penny. He just needed to make sure as many people read his paper as possible to get his ad revenue up. And how could he ensure that? By making the newspaper as attention-grabbing as possible through brain rot.

Crimes, scandals, cannibalism, tragic suicides: all had a place in *The Sun*. As Professor Tim Wu of Columbia wrote in his book *The Attention Merchants* (2016), Day made his paper "alluring to the broadest segment of society—by any means necessary" and found "his best stories at New York's police court" by reporting on its endless parade of drunkards, wife-beaters, and conmen.

Despite this dogged journalism, the story that really sky-rocketed *The Sun*'s circulation was entirely made up. It was a six-part series titled "Great Astronomical Discoveries Lately Made by Sir John Herschel, L.L.D." Apparently, Sir Herschel had developed a super-powered telescope that allowed him to see the surface of the moon in extreme detail. With it he saw man-bats (furry, flying four-foot-tall humanoids), blue goat-unicorns, and a giant sapphire temple. These shocking "discoveries" caused people to swarm newsstands: the series more than doubled *The Sun*'s circulation within just a few days, making it the top-selling daily paper in the world.

This was an extreme story, an exception more so than the rule, and perhaps enabled by a naive audience. But it reveals the incentive structure of advertising-based entertainment. The sole driver of financial success is ability to capture attention to sell ads; value to consumers is irrelevant. The penny press was the grandfather of today's brain rot, as it established the brain rot business model. Yet despite its early exploits, it was still confined to the corner of the newspaper stand and the physical limitations of its medium—cheap black-and-white paper that you'd flick through in minutes and toss until the next edition.

Stage 3—Broadcast Media
the 20th century

The physical limitations of brain rot didn't last for long, however, because the invention of the telegram and radio soon allowed instantaneous broadcasts across the globe. There were radio stories every night that people would tune into from home, including the immensely popular minstrel show *Amos 'n' Andy* in the 1930s.

Radio was soon overshadowed by television, which added visuals to radio's audio. TV also ran advertisements, meaning the shows that maximized viewers made the most money, necessitating the same dynamics as the penny press. Hence the rise of shows like *Playboy After Dark* with Hugh Hefner, *The Jerry Springer Show* with guests fighting over cheating, incest, and other gossip, and *The Benny Hill Show* in the UK, which was full of slapstick sketches and sexual innuendo. Now people could watch low-brow recorded images on a screen, alone, for hours every day. And that's exactly what they did. From the 1950s, TV skyrocketed in popularity until its peak cultural supremacy in the 80s and early 90s. Cultural critics lampooned the excessive screen time and degeneration of culture. Was there such a thing as too much entertainment? Too much drooling over screens? The fear of brain rot started to permeate the public psyche.

One such critic was Neil Postman, who wrote in *Amusing Ourselves to Death* (1985): "When a population becomes distracted by trivia, when cultural life is redefined as a perpetual round of entertainments, when serious public conversation becomes a form of baby-talk, when, in short, a people become an audience, and their public business a vaudeville act, then a nation finds itself at risk; culture-death is a clear possibility." Postman argued that TV turned everything into entertainment: news, politics, religion, education, even war, all packaged to capture as much attention from the viewer as possible. While a print culture of books, newspapers, and speeches supported logical, sequential thought, TV culture supported ephemera, emotions, and appearances above everything.

David Foster Wallace played with a similar idea in his novel *Infinite Jest* (1996). In the book, a film of the same title (nicknamed "the entertainment") is the most powerful weapon on the planet. It's so entertaining that once you start watching, you'll never stop. One of its first victims sits in a special electronic recliner with a food tray strapped to his head, allowing him to eat without moving his eyes. He soils himself while watching the film in a recursive loop, unable to even walk to the bathroom, and will stay that way until his death or someone discovers him and intervenes. *Infinite Jest* is so addictive it can be deployed as a weapon: just give your enemies the movie and they'll entertain themselves to death.

In the real world, the average household TV was on for over seven hours per day in the 1990s, according to Nielsen

Media Research. Since there were usually multiple people in each household, often alternating their viewing time or leaving it on in the background, the average per person must have been lower, perhaps around three hours. The time-sink alarmed many, but it was tolerable and life went on. While brain rot had penetrated the home, at least it was confined to a clunky box in the living room and scheduled programming. We couldn't take it to work or the bathroom, and if we wanted to watch a show we'd have to tune in at a certain time and place. I believe this semi-tolerability also normalized brain rot behavior among Boomers and Gen X, the parents of Millennials and Gen Z, so they were not as alarmed by the next, final stage of brain rot's evolution. Three hours per day was not great. But those are now rookie numbers for media consumption.

Final Stage—The Algorithms
the 21st Century

As brilliant as Postman and Wallace were, even they could not foresee what would come next: the nuclear bomb of brain rot, something so devilishly entertaining it made all previous dominant mediums look like child's play. An invention that changed everything, and printed *trillions* of dollars for its owners.

The first ingredient of this invention was the rise of the internet, where over the course of roughly twenty years, endless content was uploaded to a web accessible to anyone on the planet. Media was no longer gatekept by Hollywood executives or publishing houses—anyone could create and share entertainment, with everyone else, for free.

The problem was that also meant tons of low-quality, boring junk to sift through. The web became a maze of home videos, blogs, low-res graphics and Comic Sans. In 1996, Yahoo! was the top search engine and used a team of 20 human editors to manually catalogue 200k pages—a small fraction of the web. They were so overwhelmed with site submissions that the queue to get listed stretched for months. To find entertainment, you'd have to search their laggy directory of hyperlinks and hope for the best.

The second ingredient was the spread of laptops, TVs, and smartphones, all connected to the internet. Their screens provided much higher resolution, with better video and audio, than any TV of the 20th century. And they were extremely portable, with long-lasting batteries, meaning they could invade every bathroom, cubicle, sidewalk, and subway station in the world.

Then came the final ingredient that triggered brain rot's nuclear fission: content recommendation algorithms. These were the solution to the problem of too much boring junk online. Humans couldn't sort through the endless oceans of content, but AI machine learning algorithms could. Once unleashed, they started spidering through the internet's trillions of gigabytes of data at inhuman speeds to find the supreme entertainment hidden within.

Algorithmic Entertainment

Unlike prior mediums, the algorithms are adaptive and personalize entertainment to fit each of us individually. They analyze our browsing histories, likes, comments, click-through rates, watch times, and demographic factors to determine our exact psychological proclivities. Then they sift through an entire planet's worth of entertainment to find content that will captivate us at statistically maximal levels. We humans no longer need to search for entertainment, the entertainment searches for us.

You're already familiar with the algorithms: they hide in plain sight. They govern infinite scroll, autoplay, and personalized home pages, as well as the order of search results, comment sections, and thumbnails. Content recommendation algorithms now run every popular entertainment platform: social media, forums, YouTube, X, Netflix, most online news feeds. But you probably underestimate their existential impact. These algorithms mean we can now consume entertainment *forever,* and *the longer we consume the more entertaining it becomes.*

While hypnotic in many forms—whether text, photos, or audio—the most concentrated essence of algorithmic entertainment is the short-form video feed, popularized by TikTok in 2018. Its dark brilliance lies in requiring the least possible effort from the user: just finger swipes from snippet to snippet of content as algorithmic signals. All online entertainment platforms are converging on this sub-medium.

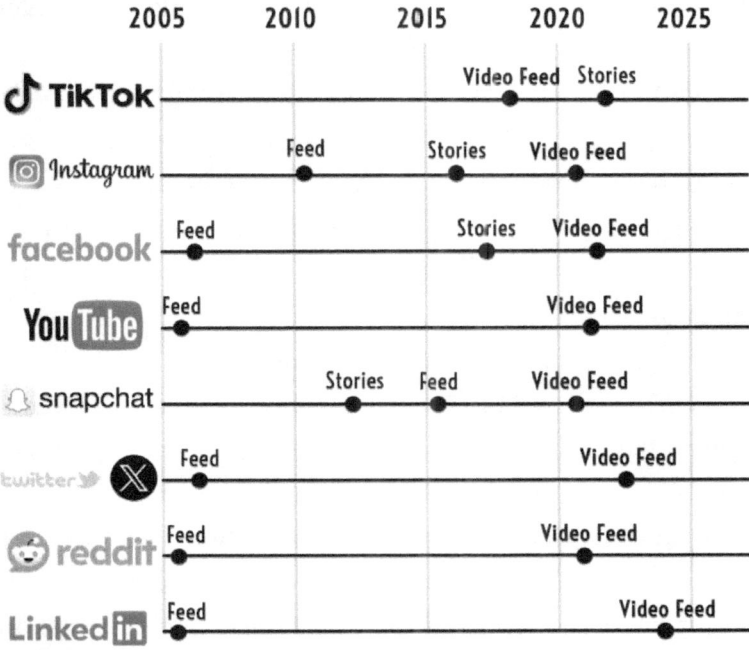

After TikTok unleashed the power of the short-form video feed, all the other algorithmic platforms hurried to do the same. It's a race to capture as much of our attention as possible, since more viewing time means more ad revenue. This incentive is evident in the earnings calls of companies like Meta (Q1 2025: "improvements to our recommendation systems have led to a 7% increase in time spent on Facebook, 6% increase on Instagram"), Alphabet (Q4 2024: "turning to YouTube, we saw robust revenue growth backed by continued growth in watch time") and Tencent (Q1 2025: "improved content recommendation which boosts user time spent and thus ad revenue").

Given this incentive, what content do the algorithms naturally promote to capture our attention?

Unfortunately, the most hyper-stimulating, mutant brain rot to ever grace planet earth: animatronic heads screaming from toilet bowls, gas station robberies, screeching game shows, war footage of drones bombing a village, virulent racism; all contextless and smashed together for maximum attention-grab.

And soon that will just be the tame stuff.

With the rise of AI image-, video-, and storyboard-generating models, the cost of creating brain rot approaches zero. Already this "AI slop" is flooding everyone's feeds, distinct in its weirdness and low quality. But it won't just be "slop" for long. The AI models have been improving at an exponential pace, developing fluency in esoteric animation styles, realism, and complex movement. Now you can prompt one and get a crazy video within minutes, ready for upload to the feeds.

Hence the "food eating food" trend, with over 100 million views, where semi-disgusting, semi-adorable creatures made of, say, bread, feast upon the bread they are made of. Or the shocking videos of plane crashes, dramatic police arrests, a woman bringing her "emotional support kangaroo" on a plane, and steroidal bodybuilders blown up to ridiculous proportions stomping around like King Kong.

AI generation has untethered brain rot from the limits of physics, reality, production budgets, or even human imagination. Things are getting weird, fast.

Even so, brain rot content on its own—whether AI-slop or not—is not the enemy. It's a symptom of the true villain: the algorithms that serve it. In fact, the algorithms *necessitate* brain rot as their end state because it is most effective at sucking up our time. The most attention-grabbing content for the human brain is rapid doses of violent, sexual, rage-inducing or otherwise startling stimuli. So the algorithms favor its spread in a sort of turbo-charged natural selection. Former Google employee Tristan Harris calls it the "race to the bottom of the brain stem to extract attention." Even Meta CEO Mark Zuckerberg has implicitly acknowledged this, stating in a Facebook note in 2018 that "left unchecked, people will engage disproportionately with more sensationalist and provocative content," in effect blaming the user, not the algorithm that recommends the content to maximize said engagement.

Content recommendation algorithms on entertainment platforms are our true enemy. They are the puppet masters who hypnotize us with brain rot. Throughout this book I often use the shorthand "algorithmic entertainment" or just "the algorithms" to describe them.

Brain rot delivery has reached its final form. The algorithms have done something no other medium has ever been able to do. We can now entertain ourselves to death, from cradle to grave, with no natural stop-signals of boredom, effort on our parts, or physical limitation. We're not even constrained by the output of human creators, because they are increasingly AIs. Humans will soon be phased out of the process entirely so that machines can devour our attention at scientifically maximal levels.

In sci-fi, there's a cliche that someday, the singularity will be reached and AI will turn evil, seize power, and enslave us, dominating us through force or working us to death.

But what if it tickles us to death?

What if we sleepwalk into a suffocating hug by AI algorithms that fill our every free moment with brain rot, preventing us from living real human lives? Life is the sum of our conscious experience over time. When the algorithms become powerful enough to harvest every free moment of this experience, that will be the true singularity. It's almost worse that we're giggling, not crying, through our enslavement.

Wallace's lethal *Infinite Jest* was far-fetched because no movie could possibly be so entertaining that we'd rewatch it over and over again. Everything gets boring after a while—it loses its novelty. But with today's algorithms, it doesn't.

There's always something new, and increasingly tailored to our interests. It is the realization of *Infinite Jest*, more powerful even than Wallace's vision. And we carry it in our pockets at all times.

Some naive contrarians (and tech lobbyists) have claimed that complaints about online media are no different from the minor hysterias caused by the type of brain rot spread by television and the written word. Just another iteration in a history of curmudgeons complaining about "the youth." And it's true, there were some complaints about previous mediums, most of which were valid. The salacious penny press, for example, or the fact people spent hours alone at home staring at their TVs every day. Those deserved at least a few curmudgeonly complaints.

But making that argument is like saying the invention of the nuclear bomb was just another advance in weapons technology, *shrug*, no big deal. People once complained about the destructive capabilities of dry powder muskets, after all, or bows and arrows before that. Why is the nuclear bomb any different? Never mind that it can incinerate millions of people and flatten entire cities at the push of a button, while dinky little bows and arrows can barely take out one person at a time.

Similarly, previous brain rot mediums were not even on the same scale of measurement. They did not provide hyperstimulating, personalized, infinite entertainment everywhere, at any time, completely unbounded by any physical constraints. There were no AI algorithms mining our attention with superhuman speed and efficiency.

Previous brain rot mediums had *end points*. We'd go to a play, watch it, then it would end. Find a book, read it, then it would end. Choose a movie, watch it, then it would end. But today, due to the algorithms, entertainment has no end. It infinitely expands: each piece of content we consume is followed by another, which we didn't ask for but is even more entertaining than the first, and then another, with each moving lower and lower on our brain stems, in ever-shorter clips, until we are yanked around in a whirlwind of ten-second jolts of brain rot, overwhelmed yet hopelessly transfixed.

And if that wasn't bad enough, there's another existentially important reason that the algorithms are different from all other mediums.

They entertain us against our will.

Enslaved Minds

How often do you quickly glance at your feeds, just to end up stuck in an hours-long binge? You decide to scroll for a couple of minutes, but once you start, something compels you to continue—against the will of your higher self. That invisible pull is the algorithms compromising your autonomy. They do this through two main forces, which I call "statistical conformity" and "evolution hacking."

Statistical Conformity

Most of us like to think of ourselves as one-of-a-kind with unique interests. But no matter how special we are, the internet has billions of users. Meaning the algorithms have a huge cache of people using their platforms who are *just like you*. Imagine them: thousands of doppelgängers—same age, same political views, same hobbies, same hairstyle—one of them is in Florida, one's in Austria, one's in Japan, and so on.

They all have hopes, dreams, and emotions just like you. And more importantly, they all have laptops and phones, and use YouTube, Instagram, and TikTok. They've watched the same videos you've watched and laughed at the same memes.

What the algorithms do is correlate your behavior with theirs to predict what content you'll find irresistible, because your doppelgängers have also consumed content that you haven't seen yet. The software engineering term for this is "collaborative filtering."

	Star Wars	Barbie Movie	Star Trek	Shrek	Doctor Who
User 1	👍		👍		👍
User 2	👍	👍		👍	
User 3	👍		👍	?	?

Consider the simple example above. Given no other data, which should the algorithm recommend next for User 3: *Shrek* or *Doctor Who*? *Doctor Who*. Because User 3 is most similar to User 1 based on past preferences, and User 1 liked *Doctor Who*. Now imagine this blown up to a massive scale: if given billions of users with billions of hours watched, you could match up eerily similar users with the same niche viewing tendencies. And this technique applies across formats—short-form video, news articles, even comment sections—anywhere engagement data is available.

Of course, this is a simplified picture. Today's algorithms are black-box, proprietary machine-learning hybrid models that layer in countless additional variables: time, location, device-type, content features, and more. But their essence lies in predicting what you will like based on the behavior of other users.

So it's not that the algorithms themselves understand what videos you find irresistible, or sense your moods, it's that they analyze feedback from people just like you. They let other humans tell them what captures their attention, then apply the statistics to you. We have little in common with AI algorithms; they don't grasp the complicated emotions and inner cravings of the human mind. But our human doppelgängers do—they have the same ones. And that's what the algorithms exploit, through the data mining of billions of people, to harvest our attention.

This also means that the way we spend our time is determined by an anonymous mass of others, not ourselves. Autonomy is derived from the Greek: *auto* (self) and *nomy* (laws). When we consume algorithmic entertainment, we are governed by the behavioral laws of others, not the self, compromising our autonomy.

Evolution Hacking

The algorithms also compromise our autonomy through content that exploits primal, evolutionary reflexes in our brains. According to neuroscientists Adam Gazzaley and Larry Rosen in their book *The Distracted Mind* (2016), stimuli that appear to offer "survival and reproductive advantages" command our attention through an ancient form of bottom-up processing that "remains deeply rooted in our modern brains." This causes our attention to not be "based on top-down goals, but rather driven by stimulus properties themselves." In other words, the choices of our top-down, conscious minds become irrelevant when we're exposed to evolutionarily critical stimuli.

Naturally, algorithms promote content with these stimuli since they're so attention-grabbing. But what does this look like in practice? How does online content trick our caveman brains into thinking it offers "survival and reproductive advantages?"

Sometimes, through those mesmerizing short-form food videos when we're hungry—we are of course evolutionarily primed to pay attention to food. Or perhaps puppy and cat videos, since scientists posit that our furry friends trigger protective parental instincts through their infantile features (large eyes, round faces, high foreheads). Even the most wizened at heart have a hard time looking away from an adorable puppy video, after all. But most frequently, the content seems to consist of a sort of turbo-charged social information. Popular content is almost never devoted to the world of objects—whether train routes, geologic formations, or math equations—and almost always devoted to the world of people.

We evolved as profoundly social animals in small groups, where our social savvy was critical both for survival and for wooing potential mates. Some psychologists even consider social cognition to be our default mental state: the "default mode network" in our brains—the system that automatically activates when our minds wander—shows remarkable overlap with the brain regions involved in social cognition. Our brains instinctively ponder other people and our relation to them. This makes us voracious consumers of social information, but particularly of the most evolutionarily critical types: drama, flirtation, gossip or "tea," in Gen Z parlance.

In the tight-knit communities of our evolutionary past, a screaming match could end in bloodshed, so we automatically pay attention to it over a peaceful bedtime story. Gossip and cruel comments about other people were also critically important to survival—they could involve conspiracy. And we perk up whenever we see flirtatious, sexual behavior, because it relates to reproduction.

Extreme social situations like these dominate our algorithmic feeds. Boring lectures delivered in a monotone by a professor might be more practically helpful to us in the modern world, but people twerking or fighting in a parking lot is more evolutionarily attention-grabbing—and consequently, far more likely to go viral. Sex and conflict mesmerize us on a deep level, which today manifests in talk show gossip, "YouTube drama," political screaming matches, and more.

It doesn't matter what our conscious brains want when presented with this stuff. It doesn't matter how smart or strong-willed we are, or whether our top-down, rational prefrontal cortices know it's all just shapeshifting LEDs. When we see violent, sexual, or otherwise hyper-stimulating situations on our screens, they provoke a bottom-up response from our monkey brains. And it's not just the type of situations the algorithms promote, but the type of people.

Our brains categorize certain people as more attention-worthy—an attractive potential mate, or a strong leader, for example. This is because in our past, if you pissed them off, you might either ruin your chances of sex or get your head bashed in. And if you received attention from them, you'd get a hit of dopamine and other stimulating chemicals. If you received attention from the weakling "at the bottom of the totem pole," on the other hand, you wouldn't really care. That weakling was much less relevant to your survival or reproductive potential. Studies show we pay more attention to high-status people than low-status ones and remember more details about them. Even just looking at a photo of someone we consider high-status activates brain regions involved in feelings of reward.

This might explain why influencers tend to be physically attractive, high-status people of child-rearing age. This also helps to explain why ripped guys are so overrepresented on the feeds, because, in evolutionary terms, a ripped guy draws much more attention than a puny one.

The key thing to understand is this type of appeal occurs on a subconscious level. Consciously, we know we will probably never meet the influencers in real life. But our primal brains are attracted to them and find them important. When shown extreme situations involving extreme people, our evolutionary reflexes are triggered to listen up, which again overrides our autonomy.

Escaping

So what do we do? How do we win an unwinnable game against algorithms that compromise our autonomy?

By not playing in the first place.

The algorithms can only mesmerize us once we've opened their platforms—once we've said, "I'll just scroll for a few seconds." If we don't open them at all, we retake control.

We only need to quit algorithmic entertainment to escape internet addiction. It's the technology that profits from wasting our lives for as long as possible. It brings the entropy, the endlessly entertaining feeds of brain rot—it is the problem.

When I first tested this, I was shocked, because the algorithms seemed like a niche sub-component of the internet. But without them, I found it hard to entertain myself with my devices. I'd flip through photos, refresh email, maybe text a friend or check the weather. Then two minutes later I'd get bored, put away my phone, and find something else to do.

The algorithms suck up far more time than you might expect. In their FTC antitrust trial, Meta revealed that in 2025, just 17% of user time on Facebook and 7% on Instagram is actually spent viewing content posted by "friends." The majority is spent on videos or Reels, consuming content by strangers. In other words, on social media it's the algorithms that waste our time, not old-school social networking. Quit them and you decimate your screen time.

Best of all, the algorithms exclusively show us content we did not *intend* to consume beforehand. If we had intended to consume it, we would have simply looked it up in the first place. As Nir Eyal explains in his book *Indistractable* (2019), distraction is the "drawing away of the mind" that impedes us from making progress toward the life we envision. It is the opposite of traction, which is derived from the same Latin root and consists of "the actions that draw us toward what we want in life."

If there is a specific video of Kardashian drama you want to watch, search it up and watch it. That's not technically distraction because that was your intention. If no algorithms are involved, after a few minutes the drama ends and you close the tab. But if while watching you see some Katy Perry drama in the recommended tab and click it, and then another video you didn't even know existed, and find yourself falling down a rabbit hole, that's *dis*-traction, going away from what you intended. And that's always what ends up happening because it's what the algorithms are optimized to do.

The interesting consequence of this is that if you quit algorithmic entertainment, *you don't quit anything you intended to consume.* You just quit the brain rot that you didn't know existed, yet find irresistible once you see, which will happen *ad infinitum* given the endless ocean of online content. The only way out is to reject the premise entirely.

Imagine playing chess with a supercomputer. You might be a strong player—the strongest in the world, even—but no matter what move you make, the computer is always twelve steps ahead. It will always win. And it never makes mistakes. You keep pushing, playing over and over again, thinking of new methods and approaches to come out on top, but you always end up back in the gutter. What would you do in this situation? How do you "win" when faced with inevitable loss?

By saying: "You know what, this is a stupid game anyways." And then you stand up and walk away. The computer makes beeping noises, untrained on how to deal with the situation. Then you go frolic in a field or something. Who's the winner now?

This is how we must treat the algorithms. The second we click onto their sites—onto TikTok, YouTube, Instagram, or news feeds—we're playing their game. We accept the terms of their framing. We can try and just watch one video, or check one thing, or use a 15-minute screen time limit, but due to their ability to compromise our autonomy, we'll always end up losing.

Instead, we must learn to not play their game at all. We must stand up from the algorithmic chessboard.

Part II:

The Solution

Isolating the Enemy

Now we will move from theory to practice and solve the problem. First, we must root out and isolate all the spidery algorithms that have infested our digital lives. If we don't, any that remain will start multiplying and stealing increasing amounts of our time… as they were designed to do.

Take out a piece of paper and list all the algorithmic entertainment that you use. Make two columns like the chart on the following page, with the platform on the left and all its algorithmic manifestations on the right. The following is an *example*—your list might look different based on your specific habits or if you use other platforms not mentioned here.

Platform	Algorithmic Entertainment
YouTube	Home page, recommended tab, shorts
Instagram	Main feed, explore page, reels, threads
Facebook	Main feed, stories, news, videos
TikTok	Everything
X	Main feed, comment threads, video feed
LinkedIn	Home page
Reddit	Home page, subreddit home pages, recommended posts
Netflix	Recommended movies and TV shows
Online news	Apple News, Google News, and other aggregators, along with algorithmic homepages of certain news sites
Video games	Everything **See explanation in the Edge Case section, below**
And so on...	

When determining what to put on your list, remember that none of the beneficial, practical uses of the internet involve algorithmic entertainment. Consider, for example:

- Using Google Maps for directions
- Ordering a ride (Uber or Lyft)
- Online banking
- Buying soap on Amazon
- Googling a math concept for your homework
- Applying for jobs
- And so on…

If an online service is *not* algorithmic entertainment, it falls into at least one of the two categories below:

1) **Not entertainment.** Entertainment is something you consume for amusement or stimulation, without a practical goal in mind. So it is not entertainment to have a problem, search for it, and click through Reddit comments for an answer. Same with using dating apps to set up real-life dates, in which case they're logistical tools, not entertainment.

2) **Not algorithmic.** This could include using the messaging sections of a platform, looking up one specific post on social media, or deciding to watch that movie everyone's talking about. Those are not algorithmic, but activities you intended to do, do, and then end. If you get the news directly through a newspaper or magazine, that's curated by human editors, unlike the endless feeds on Google, X, or Apple News. If someone sends you a specific article or post, that's also not algorithmic—just don't start scrolling after you click on it!

Beyond technical classifications you should have an intuitive sense of what algorithmic entertainment *feels* like. It feels very different from healthy, practical uses of the internet. It's externally directed, and feels like you're along for the ride, rather than the driver. It's chaotic and makes you feel scatterbrained. You forget yourself and your surroundings vanish, each clip is contextless, and you don't remember what you watched ten seconds before. There's a sort of hunger in your brain for more content and stimulation, with a simultaneous hum of anxiety. It's not an existentially good feeling. If you've been addicted in the past, binged for hours and regretted it afterwards, you know what this feels like in your gut. Trust this judgment when deciding what to add to your list.

Edge Case: Video Games

Video games are a unique form of algorithmic entertainment. They don't use the same content recommendation algorithms as other platforms, but they do use algorithms to perpetually adapt to your skill level, matchmake you with new opponents, and generate new map chunks and levels. Like the feeds, they ensure you'll always have something new and personalized to keep you hooked.

Games are also obviously entertainment because you don't seek any practical benefit from your play (unless you're a professional playing for money... which is ridiculously hard). This makes video gaming an immersive form of algorithmic entertainment that should be treated as such.

Single-player games may more closely resemble experiential art pieces, and have a set end point rather than being infinite like multiplayer games. But still, if you feel you have a problem, add it to the list. I quit video games years ago after playing thousands of hours and have never regretted it.

Once you have your list ready, take a mental photograph of it. This sheet represents the toxic waste that steals so much of your time and keeps you from reaching your goals. It isolates exactly what you will be cutting from your life.

Finally, you can measure success: the extent to which you avoid these algorithms, refusing to let them yank your attention around. Once you cut them out, online platforms begin to serve you. You become the user, not the product, as your devices operate as tools, not enemies designed to manipulate you. You'll avoid the negatives while keeping all the benefits the internet has to offer.

Separation

Now it's time to disentangle the algorithmic entertainment on your list from your healthy online behavior. It's as if you have a bag of multi-colored candies, and realize the red ones are poisonous, but since they're so jumbled up with the others, they're hard to see and easy to accidentally eat. You need to go through, pluck out the red ones and put them in a separate pile. Then you can feast upon the non-poisonous candies to your heart's content.

The average person checks their phone around 200 times per day and often clicks straight to the algorithms. Over the years, that adds up to hundreds of thousands of repetitions. Often it happens subconsciously as an automatic reflex that has wormed itself deep into our minds. We'll be in the bathroom and suddenly find ourselves browsing a feed without actively deciding to. If we always have shiny TikTok and Instagram icons under our thumbs, it's easy to unthinkingly tap into them. And it is impossible to escape algorithmic entertainment if we access it without even conscious awareness, so we must interrupt this cycle.

Start by cleaning out your home screens and notifications to make it harder to "accidentally" end up on the algorithms.

Delete your video game launchers, install a different internet browser: anything to shake up your routine and change the location of your go-to algorithms.

Go through each row on your list and see how many barriers you can erect to block access. For example, remove Instagram from your phone so you can only access it on your laptop (for the handful of non-algorithmic uses you have on your list), or deactivate your account and remove the app if you see no practical uses at all. You can simply delete other apps for good, such as TikTok, which is entirely algorithmic entertainment. Delete your news apps and subscribe to a magazine, newspaper, or email newsletter instead.

App and website blocking software can also be helpful as a temporary barrier against the algorithms. It doesn't address the root cause, and it's not a long-term solution, but it can initially help to interrupt the force of habit.

By following this process of separation, you prevent yourself from subconsciously, reflexively opening the algorithms. Of course, in today's hyperconnected world you can always find a way around barriers if you're determined. So now we must turn to the bigger task: freeing the conscious mind.

The Missing Key

By distancing yourself from algorithmic entertainment, you've made opening it into a conscious choice, not something done on autopilot. And with this choice comes the all-important question:

Will you *permanently* quit algorithmic entertainment?

When I posed this stark question to myself, I felt a tinge of uncertainty—a resistance to going cold turkey. I recognized the problem, yet something was stopping me from quitting… The fear of missing out.

I was afraid that by quitting, my life wouldn't be quite the same: that it would be more boring and isolated. I believed that my binge sessions benefited me in vague, yet valuable ways.

Suddenly, I realized that this belief was the missing key to understanding my internet addiction. In all this time, I had neglected Occam's razor—the principle that the simplest explanation is usually the best.

Why did I always feel a magnetic pull to algorithmic entertainment, despite my best intentions?

Because I had reasons to use it.

I had always focused on my reasons *not* to use it: how it wasted my time, ruined my productivity, and kept me cooped up in my room. But I had never considered my reasons *to* use it—what I'd miss out on if I quit.

When I examined my underlying beliefs, I found that I had four main reasons to use algorithmic entertainment:

- **Pleasure:** It feels good—my life is more pleasurable with it than without it.

- **Escape Boredom:** It cures boredom through exciting stimulation, helping me avoid the painful antsiness of doing nothing or doing tedious tasks.

- **Social Value:** It improves my social status, keeps me in the loop with friends, and gets me invited out more through social media.

- **Knowledge:** It gives me lots of new ideas, advice, and news stories: tons of free information that improves my knowledge of the world.

These were the justifications I had for the algorithms—what I thought I received in exchange for my thousands of hours spent scrolling. And they had persisted to the present day: *maybe if I just scroll for a few minutes, I can reap some benefits. Surely that's better than just sitting and doing nothing?*

Because of these potential benefits in the back of my mind, my every free moment was a seesaw battle of whether or not to binge. And sometimes, the perceived benefits outweighed the costs and I'd open the algorithms:

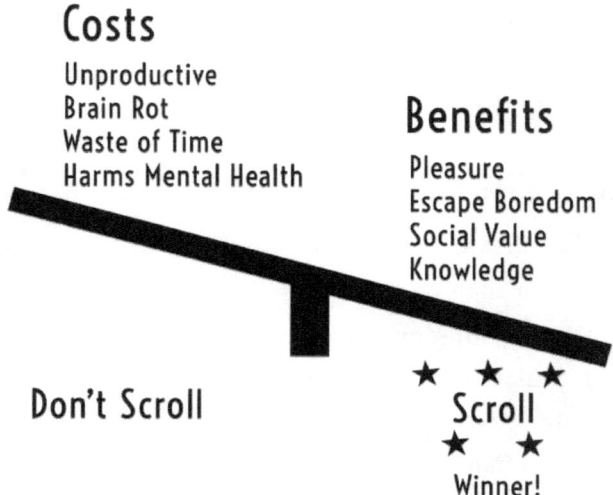

Cost-Benefit Analysis: Keep Scrolling!

There are many names for this concept: cost-benefit analysis, expected utility, cost-benefit learning, decisional balance, or even just "pros and cons." All are variations on the same idea—that our brains evaluate the positive and negative consequences of an action before deciding to do it. And studies show that even for quick or seemingly insignificant decisions—like reaching for our phones in a free moment—we still do this weighing, albeit subconsciously.

When trapped in my addiction, I had never actually made the benefits side of the scale explicit. They had always been vague perks floating nebulously in my mind. They created the uncertainty that made me resist quitting the algorithms. I didn't know what I would be missing out on, so I took the "safe" option of continuing my habits.

Yet by listing each of the benefits, I already started to feel the fear of missing out dissipate, because each one seemed a little flimsy. *Is that it?* Does algorithmic entertainment truly provide these benefits? Could I achieve them more effectively elsewhere?

Like a kid afraid of the dark, convinced a monster is hiding in his closet, the fear vanished when I turned the lights on and saw the closet was empty. All the supposed benefits of the algorithms were illogical. And these unexamined beliefs were what led to my fear of missing out, my faulty cost-benefit analysis, and ultimately my endless scrolling.

This process of illumination is exactly what we'll do in this book. We'll take the benefits and expose them as faulty, reducing their weight on the scales of your psyche.

Cost-Benefit Analysis: Stop Scrolling!

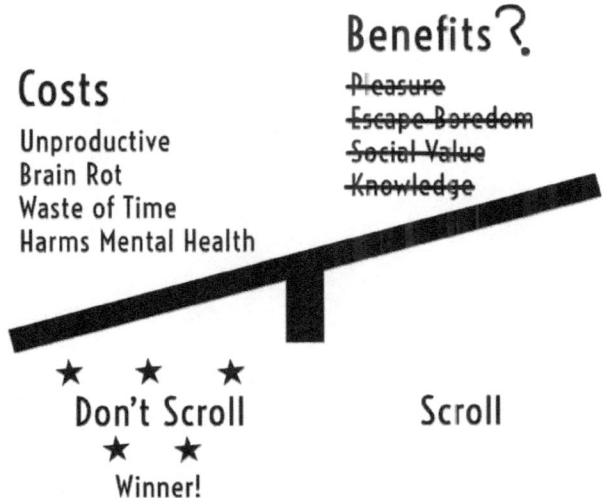

Internet addiction does not operate through physical dependence—there's no injection of a substance or withdrawal sweats that keep you coming back. It operates through psychological dependence—its way of confusing you into thinking that you need it, are missing out, or that your life is worse without it. Once you clarify and debunk the benefits, you free yourself from the algorithms once and for all. You answer the question of whether to quit with a resounding yes.

But is that really enough? Isn't it *hard* to escape internet addiction? No. It's easy once you have clarity. The reason quitting seems so hard is that, in the past, you relied on willpower.

The Willpower Method

The willpower method is the default way to quit a bad habit. You desire something but know it's not good for you, so you knuckle down and resist through sheer force of will. Imagine you're on a diet, but there's a cookie sitting on the table in front of you. Sweating with the exertion, you force yourself to resist its temptations. You pick up the cookie, smell it, maybe nibble on it, then stop yourself, tormented by all the pleasures that could have been. This torture can only go on so long before you surrender to temptation: studies show that willpower is a finite resource.

If you don't address your reasons *to* use the algorithms—the benefits side of the scale—you'll rely on the doomed willpower method. You'll feel you are missing out, making a sacrifice, and relent as soon as your reserves of willpower have been exhausted.

Before I understood this, whenever I tried to quit I'd act like some enlightened monk foregoing human pleasure and rising above through the force of my ideals. I'd see the simpletons scrolling in public and secretly crave the same—but force myself to resist—all while ticking down the hours of my "dopamine detox" as if it were some strange, foreign state of existence. The only problem was I could only stay an enlightened monk for so long until my willpower ran out and I'd rush to my room to gorge on my feeds. At no point did I forget about them or live freely.

All conventional methods of quitting implicitly employ this willpower method: screen time limits, black-and-white mode, lock boxes. Counterintuitively, these methods actually put algorithmic entertainment on a pedestal. They make it into an irresistible treat, full of charms, like a forbidden fruit. It's like when parents tell their teenager, "You can't go to that party, no way!" and succeed only in making the party seem that much more attractive. The off-limits desires are always the most enticing.

But if you debunk the benefits of the algorithms, you'll no longer have anything to resist. They will become a party you have zero desire to attend. It's only the things you desire that are hard to quit.

This simple yet powerful idea was first popularized by Allen Carr in the 80s with his *Easy Way to Stop Smoking* seminars and books. He did not admonish smokers with scare tactics such as photos of cancerous lungs or old people on ventilators. Instead he focused on debunking the faulty reasons they *desired* smoking, the supposed benefits: to relax, look cool, feel good. Quitting then becomes the "Easy Way" out as you no longer feel you're depriving yourself. This purely word-based method remains one of the most effective ways to quit smoking: some studies show that it outperforms even gold-standard clinical treatment methods.

I believe that many of Carr's ideas on willpower can be even more successful when applied to internet addiction. Smoking involves an external chemical dependency, nicotine, but with internet addiction there is no drug: it is an addiction of the mind that can be treated by the mind.

In the following chapters, we'll tick through the top benefits the algorithms purport to offer—what we'd "miss out on" by quitting—to consider their validity and whether we can get them more effectively from other sources. You'll soon find that the showy façade that the algorithms project looms far larger than the reality... "Pay no attention to that man behind the curtain!" says the algorithmic Wizard of Oz.

False Benefits

Pleasure

Algorithmic entertainment often feels pleasurable in the short-term. It tickles our reward circuits with its infinitely adaptive, evolution-hacking content. The psychiatrist Dr. Anna Lembke of Stanford University has even described our phones as "modern-day hypodermic needles, delivering digital dopamine 24/7."

The exact neurochemical makeup of screen-based pleasure in the brain is complex. Dopamine is often blamed, as it's central to addiction and numerous studies have tied it to internet use, but dopamine is less a pleasure chemical than a *pursuit* of pleasure chemical. The actual feeling of pleasure may come more from serotonin, opioids, oxytocin, or other molecules.

Given the complexity of the brain, we'll simplify things by referring to this intricate blend as a composite of "pleasure chemicals." Simply put, the algorithms release pleasure chemicals in our brains. This can lead to the natural conclusion, which I made, that the more you binge the algorithms, the more pleasurable your life will be. It's like always having a little pick-me-up pleasure button in your pocket. This then becomes a supposed benefit on the decision-making scale: the promise of a more pleasurable life.

Yet this promise is a fallacy. Because spamming your pleasure button increases your tolerance, dulling your capacity to experience pleasure in general. This is ultimately due to *homeostasis*, a fundamental law of biology that can't be cheated.

Homeostasis is the body's way of maintaining internal equilibrium despite changes in the external environment. For example, body temperature is maintained to a remarkably specific degree—usually 98.6°F. When it's hot out, we automatically sweat to cool down, and when it's cold we shiver, teeth chatter, and blood moves inwards from our extremities.

Our experience of pleasure works similarly. The brain tries to maintain its stability because it is an evolutionary signal—it helps us choose one behavior over another. If everything felt pleasurable, the feeling would not further our evolutionary goals or aid in decision making. We'd lie at home in bliss, pleasured by insignificant things and unmotivated to pursue more. But if nothing was pleasurable, we wouldn't have any motivation either—because what would be the reward for our efforts? To avoid both extremes, our brains calibrate our sensitivity to pleasure chemicals based on how frequently and intensely they're released. This can be thought of as our "pleasure tolerance."

Studies show that if you engage in lots of high-dopamine activities your dopamine receptors "downregulate" to compensate, desensitizing you to all dopamine-releasing activities. The literature often focuses on drug addictions, but one study expressly measured internet addicts and found they too had reduced dopamine receptor availability in their brains compared to non-internet addicts.

This dynamic occurs with the other pleasure chemicals as well, including opioids and serotonin—the more that are released, the more desensitized their receptors become. It's an established finding in addiction research that no matter the drug (or behavior), the more often you use it, the higher the dose you need to achieve the same effect, and the less pleasurable other activities become.

On the flip side, studies also show if you do not engage in many dopamine-releasing activities throughout the day, your receptors can "upregulate" and become more sensitive. You become relatively more stimulated by activities that release smaller amounts of dopamine. The human brain is extraordinarily plastic, constantly adjusting its baseline of pleasure to maintain balance.

Pleasure Desensitization

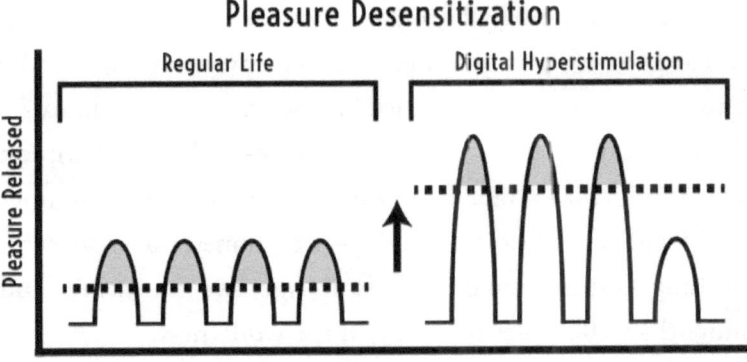

- Tolerance Level
- "Effective" Pleasure Experienced

The preceding graph illustrates the dynamic. When you engage in unnaturally high-pleasure activities, your brain raises your tolerance to the pleasure chemicals—represented by the dotted line. This stabilizes the total net "effective" pleasure that you actually feel.

The peak on the far right shows how after digital hyperstimulation, the pleasure released from a normal, healthy activity does not even register, since your tolerance is so high. What's dangerous is that this tricks you into thinking that offline life is bland, when in reality if you just give it time, your pleasure tolerance will decrease to adjust to lower levels of stimulation.

The Gamer and the Hiker

During my junior summer in high school I spent every day gaming, watching YouTube videos, and scrolling social media. I lived in a state of digital hyperstimulation. Even eating meals became excruciatingly boring if I didn't have my phone. I was so used to hits of digital pleasure that my brain adjusted by creating a super-high tolerance for the associated chemicals.

Then my parents sent me on a three-week hiking trip on the Appalachian Trail. It was organized through a summer camp with a strict no-tech policy—no phones, no wristwatch, nothing. Just me, ten other kids my age, two counselors, and more than 100 miles of hiking in the wilderness.

I was teleported from a world of digital binge sessions to nothing but staring at trees in the wilderness for hours, walking silently alongside people I didn't know. The first couple days were hell; I felt listless and complained about the sweaty hiking days and the unpleasantries of sleeping in a tent. My brain still had a high pleasure tolerance, and now that my devices were gone, nothing even registered.

But within just a couple days, my brain recalibrated. My pleasure receptors up-regulated and started picking up signals from the little things: a tin of coffee brewing on the stove, a beautiful sunset. I loved talking with the other campers, trading riddles and scary stories around the campfire. I enjoyed the splendor of the forest, the leaves fluttering in the wind, the trickling streams, and the chirping birds. I'd get genuinely excited to read the book I had packed… and I had barely read for pleasure in years. I started picking up all sorts of nuance and beauty in life that I didn't even have the capacity to notice before. My brain's reward system had adapted to the new, lower-simulation environment.

The problem was, I didn't understand this. So when I got back home, I figured getting back online would only add to my life as a net-pleasure positive. I still believed the fallacy that bingeing the internet would make my life more pleasurable. Of course it didn't, because the overstimulation soon increased my brain's tolerance to pleasure chemicals once again, regular life became boring, and I needed ever-greater digital hits to achieve the same effect.

Once you quit algorithmic entertainment, your brain's sensitivity to pleasure will increase, helping you to enjoy the little things. Your life will be just as pleasurable, and more fulfilling, than it was before.

If you never binged the algorithms in the first place, time without your devices would feel perfectly pleasant. You'd live in a state of normalcy—in other words, how people felt for all of human history until a couple decades ago. They were not in pain whenever they didn't have screens within reach—that's the illusion that today's algorithmic entertainment creates.

Summary of Fallacy #1: Pleasure

False belief: The algorithms make my life more pleasurable than it would be otherwise.

- The algorithms do trigger cheap spikes of pleasure in your brain.
- However, your brain soon adjusts to these spikes through homeostasis to prevent overstimulation, which means:
  - Your pleasure tolerance increases, making you less able to experience all other kinds of pleasure.
  - The pleasure from regular activities barely registers since you're in a deficit state. As a result, everyday life feels boring.
- Once you quit, your brain will re-sensitize and you'll find just as much pleasure in more fulfilling, real-world behaviors.

Escape Boredom

The algorithms seem to provide an escape from tedious moments: bathroom breaks, chores, downtime at work, meals spent alone, bus rides.

During these moments we weigh our immediate options: 1) use the algorithms, or 2) continue being bored. It's not that we choose them over a romantic date, an intense negotiation, or a dinner party. Most of us can peel ourselves from our screens in those scenarios. It's when we anticipate feeling bored that they become irresistible.

Imagine you're at home after work and have no plans for the evening. Everything's quiet. You flop on the couch and stare at the wall. Within seconds, you become distinctly aware of the weight of the phone in your pocket and see your black TV screen and your folded laptop nearby. This is when the algorithms magnetically pull you in. How could you resist? We all have a compulsion to fill the blank space—the terrifying void of boredom.

Yet the terrors of boredom are an illusion. In this chapter we'll explore why the seemingly empty, blank moments in our lives are actually far superior to algorithmic binges. And soon we'll lean into these moments, rather than avoid them.

Reverie

Most of the time, the state of perceived boredom that we try to avoid is not true boredom. True watching-paint-dry-boredom is hard to achieve in the 21st century. Instead, it is simply time alone with our thoughts. In today's world, with constant text pings, music in our ears, screens in every room, it's possible to spend every waking minute besieged by the thoughts of others and never spend any time with our own. Our minds become mere receptacles for online content. We bounce in a frenzy from external demand to external demand, never taking a moment to ourselves.

As a result, we perceive any time away from this constant chatter of mental intrusion as boring. Yet deep down, the mind craves this time to itself—the time to simply exist without any particular goal, to wander. The French-originated term is *reverie*, a pleasant state of daydreaming.

Reverie is when Marcel Proust in *Swann's Way* (1913) lounges at home in Paris after a long day, dips a madeleine cake in some tea, takes a bite, and suddenly feels something strange deep within: a flash of memory, from decades ago when he last tasted the flavor. It flickers and then goes dark, as he sits peacefully at home, doing nothing. Then suddenly the memory returns.

He is transported back to his childhood, in rich detail, to a time when his aunt gave him madeleines dipped in her cup of lime-flower tea:

…in that moment all the flowers in our garden and in M. Swann's park, and the water-lilies on the Vivonne and the good folk of the village and their little dwellings and the parish church and the whole of Combray and of its surroundings, taking their proper shapes and growing solid, sprang into being, town and gardens alike, from my cup of tea.

This all comes to him while sitting at home after a typical day, doing nothing but sipping some tea in a quiet room, with no distractions. Far from painful boredom, Proust experiences a blissful time for reminiscence. A time for exploring the memories and delicacies of his mind in peaceful reflection. Could Proust have done that while scrolling? Could he have pondered the beauty of his youth in full, immersive splendor while being perpetually harassed by digital stimuli?

Contrast the peaceful, Proustian reverie with the option that most choose today: the chaos of algorithmic entertainment. The turbo-charged carousel of "gyatt," "sigma," and "67,"* where we give our attention to people who flex in front of mirrors and yell political diatribes; where people will do anything just to capture our eyeballs and elicit a reaction. A place where boorishness reigns supreme because it's attention-grabbing. Where we feel drained, scattered, and guilty after a binge-session, not pleasantly enlightened like we do after a reverie-session.

---

* It's not even worth trying to explain what these "words" mean.

It is nourishing to enjoy your own company, free from the disturbances of other minds. There is a pleasant wholeness and mindfulness to it. Splintering this time with brain rot costs you something deep and important and distinctly yours.

Creativity & Strategic Thinking

Boredom is also the playground for creativity and strategic thought. In moments of reverie, we don't just sit there with "smooth brains." In fact, our brains light up with new ideas instead of just passively processing the ideas of others.

Research shows that creativity is largely a function of boredom, which is a resting state where ideas that aren't connected can play with each other in the subconscious and eventually bubble up as new connections.

One British study gave two groups of people a creativity test. One group just came in and took it as usual. The other group was forced to either read or write down numbers from a phone book for fifteen minutes before taking the test... in other words, experience excruciating boredom. When the test results came in, the "bored" group scored much higher in creativity than the non-bored group.

In his book *Digital Minimalism* (2019), Cal Newport gives numerous examples of thinkers and high achievers who relied on solitude—which he defines as time free of the input of other minds. Abraham Lincoln escaped the White House to a cabin to ponder the issues of his day, sometimes even sneaking away on horseback from his secret service detail to achieve the needed solitude. To compose his philosophy,

Nietzsche went on mountain walks and shunned all interpersonal contact. The list goes on: Newton, Locke, Schopenhauer, Kierkegaard, Woolf, Dickinson, and more. They all worked in extreme solitude. Most didn't even have spouses. They were at ease with their own company, alone, and without internet. And they were geniuses.

Consider the famous anecdotes of Archimedes discovering volume displacement when he climbed into a bathtub and shrieked "Eureka!" or Newton developing the theory of gravity by watching an apple fall from a tree. These were passive, empty moments that people today might scroll through. Consider your own peculiarly creative insights that you have in the shower—often dubbed "shower thoughts." Showers are one of the only places in the modern world where you have conscious thought untouched by screens and external influence, though many drown out this time too with music or even waterproof phone cases.

Solitude is of course a prerequisite not only for creativity, but also for its practical cousin, strategic thinking. How are you going to strategize a long-term plan for yourself while constantly distracted? How are you going to *plot* things? The supervillain needs a lair, a distant laboratory undisturbed by the harassment of other minds.

Emotional Processing

Social media—and the algorithms in general—seem to have a particularly malicious effect on mental health, especially among the chronically online youth. Jonathan Haidt's *The*

Anxious Generation (2024) and Jean Twenge's *iGen* (2017) both document the trends in riveting detail. Thanks to them, the issue is now widely recognized so I won't dedicate much space to it. Still, the damage is immense and the effects are hard to overstate.

Among US college students, rates of depression and anxiety more than doubled between 2010 and 2018, each going from around 10% to at least 20%. Suicide rates among people aged 10-24 remained stable between 2001 and 2007, then jumped by 62% between 2007 and 2021. According to practically every metric, youth mental health severely deteriorated over the same period that smartphones and social media became widespread.

I believe that this is due to ever-present distraction interrupting our natural ways of processing emotion. Throughout human history, someone would say something nasty to us, we'd have a negative thought, then we'd just kind of ruminate over it. There was no instantaneous distraction device—we'd be farming in a field or something, hoeing the soil for hours, and the sheer boredom of it would force us to process our emotions to their natural conclusions. While this can be a brutal approach short-term—to face raw, negative emotions head-on—longer-term it prevents them from festering in neglect.

Today, the second after a triggering event, we blast Radiohead and doomscroll social media to distract ourselves. The algorithms succeed in this, but this causes us to bottle up the emotion until we finally have a moment to ourselves and the unaddressed pain comes crashing down. When it does, it's vaguer, lacking a clear source since time has passed, and instead of being processed and moved on from, gradually transforms our default state into one of melancholy. We then dread time alone with our thoughts, leading us back to the algorithms for distraction, which causes us to dread our thoughts even more in a vicious spiral.

Never Bored for Long

Scrolling away every boring moment also distracts us from making any real-world plans. We distract away every free moment, then at the end of the day think all we did was avoid moments of boredom. Which is true, but if we had just let ourselves feel that boredom for a few minutes, we would have naturally found something to do—an event, hobby, new book—that could have provided much more satisfaction.

In other words, the algorithms distract us from making any new plans, which opens up more boring, empty moments, which leads to more algorithms—another vicious spiral.

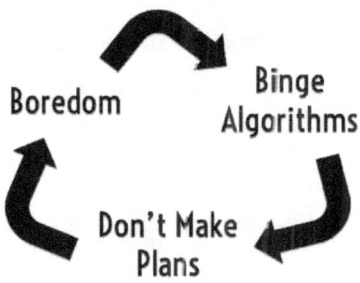

This spiral also interacts dangerously with the pleasure desensitization we explored in the previous chapter. Due to the desensitization of our pleasure-chemical receptors, whenever we take a break from the algorithms we're in a deficit state. This manifests as anhedonic laziness, an inability to find pleasure in offline activities, so we never start anything new, which in turn creates more boredom and more scrolling.

Once you quit the algorithms and snap the cycle, boredom will no longer come with a pleasure deficit; instead, it will motivate you to do new things—or that work you've been avoiding, as the alternative of staring at a wall is even less appealing. Boredom serves the valuable function of getting you outside and doing stuff. It's your brain's natural signal: you should be doing more with your life! When you allow yourself to be bored, you're never bored for long.

———

Always remember the opportunity cost. If you hadn't binged the algorithms, you might have started reading a life-changing book or joined a club and met your soulmate. Or perhaps you would've come up with a brilliant new startup idea. Or experienced deeper emotional processing, resulting in a higher sense of purpose or better mental health. Those opportunities are all impossible when sucked into a vortex of MrBeast videos. You never truly know what all that wasted time crowds out. Often it crowds out the most valuable human time you have—the time to yourself.

Summary of Fallacy #2: Escape Boredom

False belief: The algorithms "cure" boredom by letting you escape tedious, empty moments.

- In today's world, most perceived moments of boredom are just rare time alone with your thoughts.
- These moments are for reverie—daydreaming, reminiscing, and reflection—and feel pleasant compared to algorithmic chaos.
- They are also necessary for creativity, strategic thinking, and emotional processing.
- Whenever you let yourself experience boredom, you won't stay bored for long: you'll find something to do.
- Beware the vicious cycle of bingeing algorithms instead of making plans, which leads to more boring time, which leads to more algorithms.

Social Value

You might think that quitting the algorithms—particularly social media feeds—would hamper your social life. Many are billed as social networks, after all, and we are innately social creatures who fear being left out. But this fear is unfounded. In fact, the more we use the algorithms, the more our social lives suffer.

Friendship & Loneliness

In the past twenty years, we've gone from zero social media to omnipresent social media. Instagram, Facebook, X, and Snapchat now have billions of users, and it's rare for a young person not to use at least one. This means our society has undergone a massive natural experiment, one that can be evaluated at two distinct points in time: pre-social media and post-social media. So let's compare our social lives at each point.

	Pre-Social Media (2003)	Post-Social Media (2023)
% of people reporting more than four close friends	62%	38%
% of people reporting zero close friends	2%	8%
Daily time spent with friends (all Americans)	60 minutes	26 minutes
Daily time spent with friends (15-19-year-olds)	197 minutes	57 minutes

Sources: Gallup, Pew Research, American Time Use Survey (see Notes section for more detail)

These statistics are curious since social media was supposed to improve our social lives. It was supposed to help us keep in touch with old friends, send cheerful comments to each other, and meet new people. Unless… it doesn't actually work. Even after widespread adoption, social media platforms have failed to achieve their promised effects. They claimed to be big, happy friendship parties to make us feel like we'd miss out if we didn't sign up. Instead, all they did was keep us alone and scrolling.

Why Social Media Fails

Earlier in this book, I cited Meta's courtroom presentation where they revealed that just 17% of user time on Facebook and 7% on Instagram is spent viewing content posted by friends/accounts we follow (and many of these "friends" are already random distant ties we'll never see again in real life). Most time is spent on the content of strangers, served to us algorithmically. Our feeds increasingly resemble TikTok—short-form, algorithmic content—because it sucks up the most attention.

Plus, even the posts from our friends are socially useless. It's not a party, but a graveyard. Everyone posts stuff that already happened—photos of last weekend's party, trips already taken, food already consumed. The fun has been had, it died, then those involved commemorated it with tombstones on their page. This makes the feeds entirely irrelevant to planning and attending future events. No one makes friends hanging out in a graveyard, even if it advertises itself as a party. And when someone makes new plans, they don't go through their old Instagram likes and comments to decide who to invite.

Real world plans still happen almost exclusively through texts and direct messaging, not social media feeds, even for the hippest members of Gen Z. Have you ever met up with someone because you happened to scroll onto a post saying "Hey, let's meet at the grill for lunch?"

You're not missing out on any real-world events by quitting the feeds. There's no social invite waiting for you that you just have to scroll to see. That happens through text messaging or DMs.

Social Status

Most Gen Zers were introduced to social media around puberty—right when we start to hyper-fixate on social hierarchies. And our impressionable minds picked up on a pattern: popular kids seemed to have impressive profiles with tons of followers, while unpopular kids had unimpressive ones with fewer. So we naturally concluded that the cooler we made our profiles, the more popular we'd become.

However, this is a textbook causation fallacy. While social media clout is *correlated* with popularity, it usually doesn't *cause* it. People who are popular in real life tend to be more extroverted and make more friends, which translates to more followers on social media (after the fact). They also tend to care more about their image, so they invest more time on their feeds. A popular kid is popular in real life, and this causes their inflated social media presence, not the other way around.

Consider the following thought experiment:

- The sour, annoying kid suddenly gains 3,000 followers overnight on his account and his posts start getting hundreds of likes. His personality stays the same.
- The popular, fun kid's account inexplicably vanishes, leaving him with no social media presence. His personality also stays the same.

Would the first kid suddenly become the prom king and asked to sit at the cool kids' lunch table? *Hey man, we saw you have 3,000 followers now, let's hang out.*

On the flip side, would the popular kid who suddenly has no social media be shunned? *Hey man, we can't find you on Insta so we don't want to hang with you anymore.* Of course not. If anything it would make him seem more mysterious and rebellious. Some might assume he has one of those artsy accounts with only a first name or nickname listed, making it hard to find.

There are several reasons why social media profiles are not as impactful as they seem. First, studies show that the human brain is bad at comprehending large numbers. You can precisely imagine 3 bananas on a table, but can you precisely imagine 85? Yet large numbers are the only metric of the social media hierarchy: 200 likes vs 400 likes, 600 followers vs 3,000 followers. We can understand the difference, of course, and consider 400 likes to be better than 200, but it's an abstraction and not visceral.

Second, we have evolved over millions of years to be incredibly adept at evaluating social cues, expressions, physical presence, and real-life dominance hierarchies. Social media has been around for 15 years—it's a tiny bit of icing on top of the real indicators of sociality. If you have 50 followers and post blurry photos, but in real life have good posture, smile at people, and have a firm handshake, you will be more charismatic than someone who is nervous, slouches, and mumbles, even if they have 5,000 followers and post epic pics. That's why "catfishing" (creating a fake online identity) never ends well: real world presence beats online presence every time.

Besides, vanishingly few people are actually *winning* the game of social media: only the influencers with thousands of followers and sponsorship deals.

The rest of us are *losing* the game of social media and, like the chess player taking on a supercomputer, would be better off not playing at all. Because while we have nothing to gain, that doesn't mean we have nothing to lose. If we post awkward photos or political takes, appear too insecure or vain, we can scare people away. This just adds stress to our lives and overthinking for no upside. Disabling your account frees up time and energy to focus on real world socializing.

All that being said, if you've evaluated the pros and cons and do decide to keep your social media accounts, that doesn't necessitate using the algorithms. If you avoid the feeds, you can still use DMs or look up specific profiles—none of which is algorithmic entertainment. You can even keep posting yourself, just never scroll. However, this is harder than it seems because of the ease of swiping to the feed, and the fact that algorithmic homepages pop up when you first open the apps. At the very least, it is best to remove the apps from your phone and only access them on your laptop when needed.

Video Games

For many, gaming is a social experience. It certainly was for me in my middle and high school cliques, yet this gaming tended to be more "parallel play" than true bonding time. Everyone just controlled their own characters on their own screens and muttered reactive "oh no's" or other commentary into the mics. No one really spoke *with* each other, but *at* each other. Conversation centered on gameplay logistics: "Go top!" or "Make sure to buy this potion."

Socialization has a quality axis, not just quantity. That's why the movies make for such a bad first date. Shared media consumption is not the same as shared conversation.

If you are in one of these gaming friend groups, I suggest transitioning offline for an exponentially richer experience. I'm convinced a 30-minute coffee with your gamer friend face-to-face will improve your relationship more than a 5-hour gaming marathon. Or, if you live far apart, try a voice chat with no gaming involved. If you still crave playing games with the group, tabletop games provide an amazing real-world replacement.

Texting

Texting is so crucial to socializing today that it's worth considering in more depth. First of all, texting itself is not algorithmic. At its core it's a digital version of physical mail, not so far removed from the telegram. It's a holdover from the plastic flip phone keyboard: *Meet u there at 6:30. K?* So despite the false equivalency that pop culture suggests, rebelling against the attention economy does not necessarily mean quitting texting or being unreachably off-the-grid.

However, with the dominance of brain rot in today's culture, texting culture has devolved. Conversations increasingly revolve around content someone stumbles upon during their own algorithmic binges. Developers have now added preview windows to facilitate this: we can watch YouTube videos or read tweets from the confines of iMessage.

We all know the feeling of receiving these memes or articles. We look at them, crack a tiny grin, then send a *laugh* or *love* reaction. They are almost always funnier to the sender than the recipient. And it's a strange form of socialization, given the sender didn't create the message they sent, nor did you create the response when you used the built-in reaction.

If one of your friends who constantly sends you this stuff stopped, and instead just texted their own jokes and observations, would you think less of them? No, you'd probably think more of them as you'd be privy to their actual personality.

While innocently intended, brain rot texts serve as gateways for the algorithms if we click into them and keep scrolling. The preview window is actually helpful in that regard—we can see what someone sent without getting caught up in the feed.

In other words, anything a human sends you is fair game to consume and react to. Just don't send them algorithmic entertainment yourself, because that necessitates bingeing the feeds.

If you can avoid the algorithms, checking texts will start to feel like the mundane act of checking your physical mailbox: occasional, mostly logistical. In theory, your physical mailbox should actually be more stressful than your digital mailbox because a lot of bills, jury summons, and other important docs are dropped in it, while your texts are mostly just banter and social invites.

This chapter aimed to neutralize the fear that quitting the algorithms means missing out socially. But the best way to accomplish this is to live it, and realize for yourself that far from missing out, you will become more social and start going out more. In fact, almost no one will even notice you have quit, just that you seem more put together, present and charismatic as you prioritize socializing in the real world.

Summary of Fallacy #3: Social Value

False belief: The algorithms make you more popular; quitting them would harm your social life.

- People have fewer friends today than they did pre-social media.
- Social media does not facilitate real-world hangouts—it's a graveyard of past experiences, often from people you barely know.
- No one makes plans by scrolling algorithmic feeds.
- It is difficult to improve social status and popularity through social media—the causation goes the other way around.
- However, it is possible to *harm* your social status by committing faux pas, and people are judgmental. Why not play to your strengths in the real world?
- Humans have evolved as social animals over millions of years: digits on a screen are dwarfed by real-life presence.
- If gaming is a social outlet, try voice calls, real-life meetups, or board game nights instead for a richer experience.
- Texting is not algorithmic entertainment—fixing your internet addiction doesn't mean ghosting everyone.

Knowledge

The algorithms flood us with endlessly captivating content, so it's easy to assume they make us more knowledgeable about the world. We fear that without them, we'd miss out on valuable new ideas, have no outlet for our curiosity, and end up as information-deprived rubes. But once again, the opposite is true.

Information Scarcity

We humans are "information foragers." We naturally seek out new information because in prehistoric times it was a rare commodity. We'd acquire whatever we could about hunting strategies, rival tribes, arrowhead shapes—oh, and is that bright red mushroom poisonous? An irresistible urge to learn was evolutionarily crucial given that we relied on brains over brawn. The optimal strategy was to suck up every bit of information, no matter how seemingly irrelevant. We had to be vacuum cleaners, because in our nomadic tribes, there wasn't much to go around.

The problem? Today we have access to infinite hyperstimulating information, everywhere. We have an abundance, not a scarcity. Every day, there are over 20 million videos uploaded to YouTube, 500 million posts on X, and an estimated 270 million posts on TikTok. So if we keep the information scarcity mindset, we will burn out by scrolling night and day, consuming content generated by billions of people. We can never see it all.

If the optimal strategy were still to consume as much info as possible, we should all be much smarter and more accomplished by now. We went from information scarcity to overflowing abundance. Now six-year-olds have iPads with access to all the world's information, plus infinite quantities of AI-generated content. There's more information than ever. But have we gotten smarter or more knowledgeable as a result?

Intelligence & Achievement

The short answer is no. In fact, we seem to be getting dumber. As in the social chapter, causality is hard to prove, but here are a few concerning trends in mental clarity that *just so happened* to occur over the past thirty years, and particularly over the past fifteen, alongside the rise of the algorithms.

The "Flynn Effect" describes how average raw IQ scores rose throughout the 20th century, increasing by about 3 points each decade. As a result, each new generation outperformed the previous by roughly 10 points (and some even peg the number at 15 points on tests of fluid intelligence). Researchers speculate that this may have been due to better nutrition, more intelligence-based labor vs manual labor, and better education. All signs pointed to this trend continuing into the 21st century.

Inexplicably, however, the Flynn Effect has been *reversing* since the late 1990s. The decline has been particularly pronounced in the youth and in the U.S. and Europe, where studies have reported drops of 2–4 IQ points per decade (or about 5-7 points below the projected historical trendline). Genetics cannot explain this shift: one Norwegian study found that among siblings, younger siblings tended to score lower than older ones (after controlling for birth order effects), despite their shared genes. In other words, something environmental has changed in recent years that appears to be impairing the minds of the youth.

This trend is curious given the narrative that omnipresent internet should make tech-savvy kids smarter by giving them access to the world's information. Unless, perhaps, their brains are fried from the constant distraction. Brain rot might indeed be a literal phenomenon. Imagine two people of equal innate intelligence sitting in the waiting room of an IQ testing center: one scrolling the hysterical news cycle, watching TikTok, and being interrupted by texts every few seconds, while the other just calmly sits, waiting, doing nothing but mentally preparing for the test. Would they perform the same?

And it's not just IQ. Even with access to classroom laptops, AI chatbots, and more education funding, kids today are performing worse academically.

From 2003 to 2022, the average scores recorded by PISA (the Programme for International Student Assessment) have marched downwards, with the trend accelerating in the mid

2010s. Given that this is an average of dozens of OECD (developed, high-income) countries, it's hard to pin the blame on any particular educational policy.

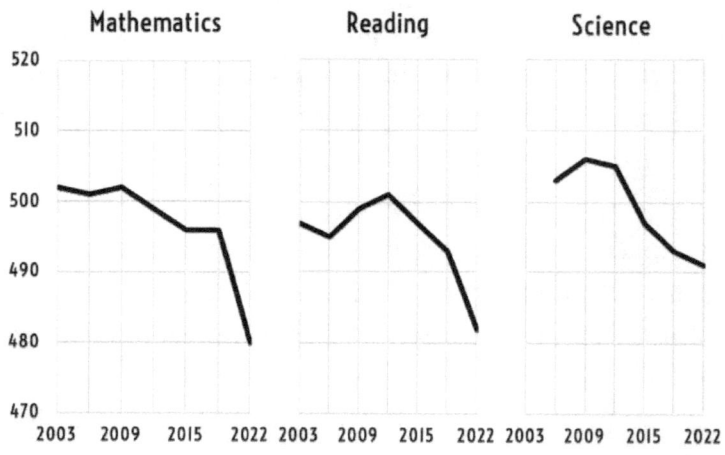

Source: OECD (2023), PISA 2022 Results (Volume I): The State of Learning and Equity in Education.

This is not just a quirk of PISA scores—in the American NAEP assessments, reading and math scores are at 30-year lows. Between 2020 and 2023, the average math score for 13-year-olds dropped 9 points, the steepest drop ever recorded, coinciding with the shift to fully remote online schooling during the COVID pandemic. Yet by the logic of techno-optimists, remote schooling should have been an educational boon. During this period, kids were distanced from the influence of technologically obsolete teachers and plugged right into the internet, the source of infinite information—and browsed it for record hours. If scrolling improved the acquisition of knowledge, they should have started performing better than ever. Yet the opposite happened.

Finally, if the external measures were not convincing enough, we can look at self-reported measures. The University of Michigan has administered the Monitoring the Future survey since the 80s where it has asked thousands of 12th graders questions like "During the last 30 days, on how many days did you have difficulty thinking or concentrating?" and the same for "trouble learning new things." The results show an all-familiar spike in the past ten years, coinciding with the reign of the algorithms.

Share of 18-year-olds who reported the following issues on at least 6 of the past 30 days

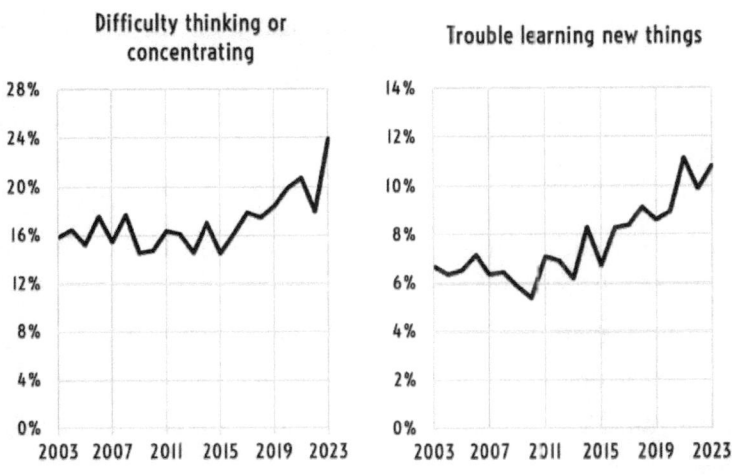

Difficulty thinking or concentrating

Trouble learning new things

Information vs. Understanding

It seems today's youth are performing worse across the board. So why does access to infinite, free information not translate to higher intellectual achievement?

First, because information is not the same as understanding. If you were to read a phone book and memorize every number in it, you'd stuff your brain with information, but your understanding would stay the same—your overall thinking not enlightened.

In *How to Read a Book* (1972), Mortimer Adler and Charles Van Doren explain that for a reader to gain in understanding, the writing must be puzzling at first. You have to ponder the ideas until you achieve the *aha!* moment of new understanding. On the other hand, "if the book is completely intelligible to you from start to finish, then the author and you are as two minds in the same mold. The symbols on the page merely express the common understanding you had before you met."

When everything in front of you is instantly digestible, you're just adding new information to your brain, not new understanding. And almost everything the algorithms show us is instantly digestible: if it wasn't, we'd scroll past it. Initially confusing, denser, puzzling information that requires work to understand is how we get smarter, but it can't compete on the feeds.

Not all information is created equal. Given their incentive structure, the algorithms naturally pump us with brain rot: political drama, sports news, and celebrity gossip. Perhaps there is an occasional two-minute, oversimplified, animation-filled explanation of the theory of relativity to create a "mind-blowing" moment, but never the long lectures required to truly understand it. The feeds merely stuff our brains with line after line in the phone book of salacious gossip and outrage.

Working Memory

The second problem with algorithmic information is our physical limitations. We rarely commit the information we consume to memory. This is because our working memory—the brain's temporary staging ground for new information—gets maxed out. As Nicholas Carr explains in *The Shallows* (2010), "working memory is the mind's scratch pad" whereas "long-term memory is its filing system." And this scratch pad can only hold a few items at a time: one 1956 study suggested around seven pieces of information, though current evidence pegs the number at around two to four. In psychology, this is called the "cognitive load."

Carr writes that when our cognitive loads are too high, we become "unable to retain the information or to draw connections with the information already stored in our long-term memory... Our ability to learn suffers, and our understanding remains shallow." He argues that in today's world of information overload, where you see thousands of bits of context-less information per day, have multiple tabs open, and infinitely scroll, it's easy to max out your cognitive load. Like a computer RAM card that has reached capacity, any additional data won't get processed.

I think this is why after you binge the algorithms for a few hours, and then try to remember what you saw afterwards, it feels like a fuzzy, forgettable haze. You might be able to recollect one or two specific pieces of content, but of the thousands of posts, videos, and comments you scrolled through, it's negligible. The content seems interesting in the moment but does not get committed to memory. That would be too cognitively demanding to do for thousands of pieces of content per day. Bombarding yourself with information leaves you less able to retain it.

The News

You might rely on the algorithms for news, and believe that they're the best way to stay informed on current events and politics. And yes, an X or Google News feed is great for bite-sized, minute-to-minute updates on breaking news, for example. But let's take a step back and ask how valuable reading this breaking news is in the first place. Does learning it in two minutes instead of in the weekend paper make any tangible difference to your life?

When scrolling headlines, ask yourself: but how does this change *my* actions? Reading the news often affects us emotionally, but do we actually *do* anything about it? It often consists of strange, clickbait-y anecdotes from across the world with no practical value.

Neil Postman observed this 30 years ago in his book *Amusing Ourselves to Death* (1985):

How often does it occur that information provided you on morning radio or television, or in the morning newspaper, causes you to alter your plans for the day, or to take some action you would not otherwise have taken, or provides insight into some problem you are required to solve? For most of us, news of the weather will sometimes have such consequences; for investors, news of the stock market; perhaps an occasional story about a crime will do it, if by chance the crime occurred near where you live or involved someone you know. But most of our daily news is inert, consisting of information that gives us something to talk about but cannot lead to any meaningful action.

Even these potential uses that Postman identifies—the weather, investing, nearby crime—have been replaced by apps (weather apps, stocks apps, or those loud buzzing safety alerts that governments push to our phones).

My solution to staying informed without incurring the brain rot of online news is to get news the old-fashioned way: from physical newspapers or magazines. These are curated by human editors, not algorithms, meaning they're the same for whoever reads them, and best of all, they're *finite*.

It only takes an hour or so to get the gist of *The Economist*, which I get weekly and read over coffee, unlike algorithmic news that sucks up every bathroom break and nighttime scroll session, and emits panicked notifications 24/7. Reading

the paper or a magazine is a pleasant, almost relaxing experience, not doomscrolling, and tends to provide a nuanced view
of the world.

If you don't like *The Economist*, you might choose *Time* or
The Week for a punchy weekly summary of current events, *The
Atlantic* for in-depth analysis of politics and culture, *The New
Yorker* for a literary and investigative slant. There are countless
newspapers and magazines to choose from. The point is about
the medium, not the politics (though of course, medium
influences politics because the algorithms promote the most
attention-grabbing positions). It's up to you whether to get a
physical copy of *The Nation* or *The National Review*; both are
better than their algorithmic equivalent.

Since starting this habit, I've never felt left out in any politics or news-based discussion. In fact, I've found that reading
the more nuanced articles rather than scanning clickbait
headlines on X allows me to consider issues on a much deeper
level. Rather than seeming like a shut-in hermit, I often seem
more well-versed in current events, all thanks to my little
weekly magazine.

You've Done It All Already

In today's world of information abundance, the optimal strategy has changed. Now the key is to be hyper-selective of the
content you consume. Choose books over threads on X, and
be selective with the books too. Only seek information when
you have a topic in mind or question you are trying to answer,
rather than letting the information come to you via algo-

rithms. Because if you do the latter, you'll be swamped with irrelevancy.

Most people operate with an information-scarcity mindset, like someone scrabbling for as many scraps of food as possible with their bare hands, even though it's a gigantic, resplendent feast. They don't realize they'll soon overeat and miss the really good stuff. Once you adopt a selective mindset, you'll have a serious competitive advantage.

In case you had any fear remaining of losing out on knowledge, remember that you've already binged algorithms for thousands of hours thus far. If the algorithms held the secret keys to success, wouldn't you have found them by now? You've probably already consumed more information than most humans in history by a factor of ten. What has all that information truly given you? Spending thousands more hours on the algorithms is not the solution. It's time to change strategies.

Summary of Fallacy #4: Knowledge

False belief: The algorithms make me more knowledgeable and well-informed than I would be otherwise.

- Information is no longer scarce but infinitely abundant—the optimal strategy is now hyper-selectivity.

- The average person has access to infinite information today yet performs worse on measures of intelligence and achievement.

- More information does not automatically equal more understanding: think of the difference between memorizing a phone book and hearing one profound, succinct idea.

- The algorithms overload your working memory, making you unable to convert memories from short-term to long-term.

- News feeds are not practically actionable: when was the last time you saw news that impacted your daily life?

- You've already binged thousands of hours of online content— if it held secrets to success you would know by now.

Summary

In the preceding chapters we have examined the benefits that algorithmic entertainment purports to offer. We've shone a light on the vague fears of missing out that kept you from quitting.

When I first did this, I felt cheated. I realized I had wasted thousands of hours of my youth chasing benefits that were never truly attainable. I received nothing in return—nothing I couldn't have gotten far more effectively from other sources.

That's when I decided to quit algorithmic entertainment for good and never consciously reopen the feeds on my list. In fact, I'd feel like a chump if I did. I saw them for what they were: parasitic brain rot machines that added nothing to my life.

It is crucial for you to believe that there are no legitimate reasons—practical or emotional—for using the algorithms. It's not worth it! You should *know* this with certainty, not trick yourself into thinking it. If you have doubts, I recommend revisiting the relevant chapter and reading the summary bullets. And if you have any other reasons to use them not covered in this book, be sure to prod their validity on your own.

Due to their intentionless, externally directed nature, the algorithms are absurdly inefficient at helping you achieve internally directed goals. Any flimsy benefit they offer can be more effectively derived elsewhere. They are designed to financially exploit your attention, nothing more.

Once you've considered your reasons *to* use them more carefully, quitting becomes the easiest decision in the world. There are no more pretexts to lure you back. No tempting benefits to offer; no weight on the opposite side of the scale. Just nightmarish chaos that leads to anxiety, pleasure deficit, lack of adventure, loneliness, and cognitive overload. Why would you ever desire that? This means you no longer need the willpower method but instead can quit with what I call the "Apathy Method."

The Apathy Method

Having dispelled the nebulous benefits, you realize that quitting the algorithms can't hurt you. You stop investing in them and see them for what they truly are: wholly irrelevant noise. Clowns and monkeys dancing at a circus, desperate to sell your attention for ad revenue, but fundamentally pointless.

You no longer have to struggle against your urges, because what would you be struggling against? The algorithms don't offer you anything that you have to wean yourself off of.

Previously, you were struggling against your temptation to eat some delicious marshmallow. Now you realize it's a charred, carcinogenic marshmallow with worms crawling around in it. You don't even need to keep yourself from eating it, because it's disgusting and gives you nothing, so why would you even desire it in the first place? You have debunked every reason to desire the marshmallow.

What's left is apathy towards algorithmic entertainment. Apathy makes the process of quitting painless and easy, because the only way to truly be free of a vice is to forget it exists. Compulsion against a desire is a compulsion also—like if I tell you not to think of a pink elephant right now. Whatever you do, keep the pink elephant out of your mind's eye, and don't think of it... Of course you will. And that's part of why the willpower method is doomed to fail.

Think of the most successful, admirable, ideal person, whom you aspire to become. Do they spend five hours a day on their feeds? No, that would not even occur to them. Do they use site blockers, black-and-white mode on their phone, or screen-time apps? Also no.

Some people in the world are *completely* free from algorithmic entertainment. They're at the pinnacle of success—valedictorians, Olympians, CEOs and scientists—because they are apathetic to it. It does not interest them, and they find it unimportant. It is a devastatingly simple truth.

What do these successful people think of all the temptations of social media, online gaming, forums, and streaming?

They don't.

Before you were introduced to algorithmic entertainment, whether as a young kid or later on, you shared their mindset. You didn't feel you were depriving yourself when running around the playground without a phone, did you? Did the worst digital vices you have today feel like temptations back then? Of course not. You were having a blast, without any crippling internet habits. You weren't interested in what your parents' PC could show you.

You have probably quit some algorithmic entertainment in the past, you just don't think of it anymore. Maybe you used to binge video games but stopped when you went off to college. Did you use Snapchat nonstop? Tumblr? Pinterest?

Or Twitch? At the time those sites were addictive, but no longer.

You somehow completely forgot about those platforms and it didn't take fist-clenching effort to do so. It just took apathy. You just don't do them anymore, it's that simple. You got tired of them, realized they had nothing to offer you, and quitting was painless.

This book helps you apply that same feeling, that same success you've had in quitting other toxic online behaviors, to all that remain. Someday you will look back at your list and think, *I just don't do that stuff anymore. I couldn't care less about it.* The brain rot that has infested your life will soon become a distant memory.

This is the only way to truly be free. If you rely entirely on an app blocker and are at constant war with your desires, or feel you are depriving yourself like some tortured poet, you'll never truly be free. That's the doomed willpower method.

You will finally be free when you realize it's been days and you haven't even thought once about those old algorithms. Because why would you?

After reading this book, I hope you feel a peace wash over you. You've removed a massive weight from your psyche. There was a tense battle in your brain, one side arguing for all the supposed benefits of the algorithms, the other admonishing you to stop. And this battle happened all day, every day, whether in your conscious or subconscious mind. You were trapped in a cycle of bingeing and regret. But now you've realized that the benefits are an illusion, and the tension is gone.

Your chess game against the supercomputer, in which you were clenching your fists and sweating with effort, is over. You realized the whole game was a fraud, calmly stood up, and walked away from the chessboard. *That's* how it feels. The struggle is over. And you'll never have to fight this battle again.

Now when you come home from a day at work, instead of instantly pulling out your devices and booting up the algorithms, you'll relax and sit, no screen in sight, and plan a real-world activity that you actually want to do. You won't be resisting the algorithms, because what is there to resist? You have punctured the false beliefs that kept you hooked.

And if you ever do find yourself mindlessly scrolling again, don't feel shame or that it's some devastating relapse. It shouldn't take shame to stop yourself—rather, just realizing that the behavior is illogical. You simply forgot for a second that the algorithms provide nothing of value, so remind yourself of the truth, and shrug it off. They have no power over you anymore. Again, they're nothing but clowns and monkeys dancing at the circus: you see them for what they are.

Beyond Brain Rot

O nce you've decided to quit algorithmic entertainment with the apathy method, you have freed yourself from internet addiction. You found the key, effortlessly opened your cell door, and are now free to walk out. So what now? What is life like, beyond brain rot? Most of us in Gen Z don't remember a time before the algorithms. Our lives have always been permeated by the hum of overstimulating, contextless digital chaos. We were tricked into thinking it was exciting and enlightening, that our lives would be drab without it. But the opposite is true. A life free from the algorithms is infinitely better: more fun, productive, social, relaxing, and existentially satisfying.

This might seem hyperbolic, like I'm promoting an algorithm-free life as some sort of panacea. But in many ways, it truly is a panacea. The algorithms corrupt our daily lived experience, displacing hours of our free time across hundreds of splintered moments throughout the day and injecting them with distraction. Removing the algorithms is therefore one of the most consequential changes you can make. The effects are difficult to overstate.

When I first quit the algorithms for good, it felt miraculous. I started noticing stunted abilities and subtleties of life that now had room to grow. You will notice these too. Life beyond brain rot helps you realize superpowers you didn't know you had.

First, you'll feel a sensation of clouds clearing from your head. You'll finally be able to hear yourself think. The whirlwind of content you consumed daily turned your thoughts into a dizzying nightmare of randomness. As Earl Nightingale once said: "we become what we think about." Now you will stop thinking about *Ballerina Cappuccina** and start thinking about uplifting, productive things.

Your headspace will stop being degraded, your working memory no longer overloaded, and your dopamine receptors will sensitize—all due to your brain's immense neuroplasticity. Your brain had been optimizing for a scattered, stressful life of digital hyper-stimulation and external demands. Now it will be optimizing for your life, ideas, and creative output.

Once you get rid of all the algorithmic noise, you'll also lose the heavy feeling that a thousand judgmental eyes are watching you on social media. You'll be able to do your own thing and be spontaneous, not stressed about what people who browse your profiles might be thinking, or getting just the right video on your story, or all the other strange neuroticisms that social media brings. And there will be fewer

---

* An AI-slop ballerina with a cappuccino cup for a head, part of the trend of surrealist "Italian brain rot" that flooded TikTok in early 2025

manic, meaningless celebrity dramas that the feeds use to exploit your attention.

After a long day at work you'll be able to truly sit and relax, and not stress yourself with even more digital stimulation and external input. You'll reflect and think things over throughout the day instead of leaving it all to the second you lay in bed, when your mind becomes a stressed whirlwind due to its neglect.

You'll also feel a curious sensation that you are in good hands: your own. You'll be able to depend on your future self more because you'll be more present. At first it's the small things: doing your laundry, showing up places on time, getting your paperwork in order. You'll start studying for the test a week in advance, not one day in advance. Your life will be so much easier when you allow yourself to have empty time to yourself where you can do your chores, take care of yourself, and plan ahead. This will provide a huge relief from the anxiety you have humming in the background as an internet addict.

And when opportunity arises, which it will, you'll have the presence of mind to jump on it proactively rather than always playing catch-up. An internet addict is always on defense, with a mindset to do the bare, acceptable minimum, and then get back online. Once you escape, you'll start going on offense, which is where the real rewards lie. Remember— fortune favors the bold!

What's more, you'll have a huge competitive advantage over all the algorithmic addicts of today's society, given the sobering extent of the problem. Everyone on the sidewalk has their head buried in their phones. Most young people at top colleges and in top jobs are addicts—they scroll forums, shop, and browse social media in the middle of lectures. Everyone in today's generation is addicted. It's just that some addicts are slightly better at cramming for tests, sufficiently pleasing the teachers, and leveraging the internet as a tool. The world is craving the rare young adults with the capacity to sit and work for long stretches of time, who are ambitious, creative, and plan ahead while everyone else is scrolling. And unlike all previous generations, you will have the power of the internet on your side, which when used correctly, with intention, is a miracle of productivity and knowledge.

Mental clarity, emotional stability, and productivity at work are all amazing superpowers. But they pale in comparison to the greatest fruits of life beyond brain rot, which become available during your free time.

Free time makes us human. It's when we can finally be ourselves. As John Keating says in the movie *Dead Poets Society* (1989), "Medicine, law, business, engineering, these are noble pursuits and necessary to sustain life. But poetry, beauty, romance, love, these are what we stay alive for." And these things to stay alive for are precisely what the algorithms spoil the most. When we would otherwise have time to ourselves—in the quiet hours of the morning or at night, the in-between space—to do what we actually want to do, to engage in uplifting, enlightening entertainment, to have a magnificent obsession, to feel something, the ever-present brain rot emerges with its carousels of content and cheap stimulation. No longer.

You will finally have the time to read books, draw, play board games, or tinker with model airplanes. You will have time with your own thoughts, to wander through a forest smelling the cedar wood, or simply watch raindrops trickle down a window, merging as they cross paths. You will have time for art, glamor, and romantic intrigue, time for a dinner party spent in the moment, enjoying each other's company, not checking your phone during every lull in the conversation, or wondering how it would look on Instagram. In short, you'll have time for all the potentialities that the algorithms have crowded out.

So stock up on books and art supplies, pick out a musical instrument, sign up for some classes or athletics; your energy and dopamine need a productive outlet, which is a beautiful thing.

When you shed the algorithms, forces will take hold inside of you. Sleeping giants will stir within—the artist, the musician, the writer, the entrepreneur. All the latent talents that the algorithms have suppressed will have room to flourish. Magnificence awaits you, beyond brain rot!

Thank You for Reading!

If you enjoyed this book, please consider leaving a quick, honest review—it truly helps others discover it and means a lot to me.

Want to stay connected? Join the *Brain Rot Resistance* for exclusive updates and early news about upcoming releases.

Sign up here → beyondbrainrot.com

Notes

The Rise of Brain Rot

10 "the 'content' of a medium": Marshall McLuhan, *Understanding Media: The Extensions of Man* (Cambridge, MA: MIT Press, 1994), 18.

12 Bibliographers estimate: Eltjo Buringh and Jan Luiten Van Zanden, "Charting the 'Rise of the West': Manuscripts and Printed Books in Europe, a Long-Term Perspective from the Sixth through Eighteenth Centuries," *The Journal of Economic History* 69, no. 2 (2009): 417, http://www.jstor.org/stable/40263962.

12 Benjamin Day realized that if he sold ads: Tim Wu, *The Attention Merchants: The Epic Scramble to Get Inside Our Heads* (New York: Alfred A. Knopf, 2016), 18, Kindle edition.

13 "alluring to the broadest segment of society": Wu, *The Attention Merchants*, 19–20.

13 "Great Astronomical Discoveries Lately Made": István Kornél Vida, "The 'Great Moon Hoax' of 1835," *Hungarian Journal of English and American Studies (HJEAS)* 18, no. 1/2 (2012): 432–35, http://www.jstor.org/stable/43488485.

13 doubled *The Sun*'s circulation within just a few days, making it the top-selling daily paper in the world: Vida, "The 'Great Moon Hoax' of 1835," 435.

15 "When a population becomes distracted by trivia": Neil Postman, *Amusing Ourselves to Death: Public Discourse in the Age of Show Business* (New York: Penguin Books, 2006), 155–56, Kindle edition.

15 One of its first victims sits in a special electronic recliner: David Foster Wallace, *Infinite Jest* (New York: Back Bay Books, 1997), 34 and 54.

15 over seven hours per day in the 1990s: Nielsen Media Research, *Report on Television: 2000* (New York: Nielsen Media Research, 2000), 14, https://www.worldradiohistory.com/Archive-Ratings-Documents/Nielsen-2000-Report-on-Television.pdf.

17 manually catalogue 200k pages: Joshua Quittner, "Seek and Ye Shall Find (Maybe)," *Wired*, May 1, 1996, https://www.wired.com/1996/05/indexweb/.

17 solution to the problem of too much boring junk online: Deepjyoti Roy and Mala Dutta, "A Systematic Review and Research Perspective on Recommender Systems," *Journal of Big Data* 9, no. 59 (2022), https://doi.org/10.1186/s40537-022-00592-5.

18 They analyze our browsing histories, likes, comments, click-through rates, watch times, and demographic factors: Francesco Ricci, Lior Rokach, and Bracha Shapira, *Recommender System Handbook*, 2nd ed. (New York: Springer, 2015), 20, 78; Amiruthaa Amudharasan, "The Impact of Recommendation Algorithms: Analyzing the Influence of Data on Marketing Strategies in the Media Sector," *Open Journal of Business and Management* 11, (2023): 3373–84, https://doi.org/10.4236/ojbm.2023.116184; Maximilian Boeker and Aleksandra Urman, "An Empirical Investigation of Personalization Factors on TikTok," in *WWW '22: Proceedings of the ACM Web Conference 2022* (New York: Association for Computing Machinery, 2022), 2298–309, https://dl.acm.org/doi/10.1145/3485447.3512102.

19 Chart on Feature Convergence Across Apps: Meta Platforms, Inc., "Meta's Opening Statement," presentation slides in *FTC v. Meta Platforms, Inc.*, No. 1:20-cv-03590-JEB (April 2025): 44, https://s3.documentcloud.org/documents/25896886/metas-opening-statement-slides.pdf.

20 Meta: Meta Investor Relations, "Meta Q1 2025 Earnings Call Transcript," April 30, 2025, https://s21.q4cdn.com/399680738/files/doc_financials/2025/q1/Transcripts/META-Q1-2025-Earnings-Call-Transcript-1.pdf.

20 Alphabet: Alphabet Investor Relations, "2024 Q4 Earnings Call Transcript," February 4, 2025, https://abc.xyz/assets/e0/99/7331ada54f6a9f65d08eeadd25d5/2024-q4-earnings-transcript.pdf.

20 **Tencent:** Seeking Alpha, "Tencent Holdings Limited (TCEHY) Q1 2025 Earnings Call Transcript," May 14, 2025, https://seekingalpha.com/article/4786684-tencent-holdings-limited-tcehy-q1-2025-earnings-call-transcript.

20 **"AI slop" is flooding everyone's feeds:** Mark Sullivan, "Enjoy 'AI slop' summer. What's coming next is worse," *Fast Company,* June 5, 2025, https://www.fastcompany.com/91346487/enjoy-ai-slop-summer-whats-coming-next-is-worse.

21 **"food eating food":** "AI Slop: Last Week Tonight with John Oliver (HBO)," YouTube video, posted by LastWeekTonight, June 23, 2025, https://www.youtube.com/watch?v=TWpg1RmzAbc.

21 **"emotional support kangaroo":** Dani Di Placido, "Emotional Support Kangaroo Video Goes Viral – But It's Completely Fake," *Forbes,* May 28, 2025, https://www.forbes.com/sites/danidiplacido/2025/05/28/emotional-support-kangaroo-video-goes-viral-but-its-completely-fake/.

21 **"race to the bottom of the brain stem":** Tristan Harris, "Persuasive Technology & Optimizing for Engagement," *Center for Humane Technology,* testimony before the U.S. Senate Committee on Commerce, Science, and Transportation, June 25, 2019, https://www.commerce.senate.gov/services/files/96e3a739-dc8d-45f1-87d7-ec70a368371d.

21 **"left unchecked, people will engage disproportionately":** Mark Zuckerberg, "A Blueprint for Content Governance and Enforcement," *Facebook Notes,* November 15, 2018, https://www.facebook.com/notes/751449002072082/.

Enslaved Minds

26 **"collaborative filtering":** Jacob Murel and Eda Kavlakoglu, "What is Collaborative Filtering?," *IBM,* March 21, 2024, https://www.ibm.com/think/topics/collaborative-filtering; Ricci, Rokach, and Shapira, *Recommender System Handbook,* 20, 77–118;

Zhenhua Dong, Zhe Wang, Jun Xu, Ruiming Tang, and Jirong Wen, "A Brief History of Recommender Systems," *arXiv preprint* arXiv:2209.01860 (2022), https://arxiv.org/abs/2209.01860.

27 hybrid models that layer in countless additional variables: Ricci, Rokach, and Shapira, *Recommender System Handbook*, 20, 191–226.

27 predicting what you will like based on the behavior of other users: Arvind Narayanan, "Understanding Social Media Recommendation Algorithms," *Knight First Amendment Institute at Columbia University,* April 26, 2023, 22–23, https://doi.org/10.7916/khdk-m460.

28 "survival and reproductive advantages": Adam Gazzaley and Larry D. Rosen, *The Distracted Mind: Ancient Brains in a High-Tech World* (Cambridge, MA: MIT Press, 2016), 64, Kindle edition.

28 "driven by stimulus properties themselves": Gazzaley and Rosen, *The Distracted Mind,* 64.

28 trigger protective parental instincts: John Archer and Soraya Monton, "Preferences for Infant Facial Features in Pet Dogs and Cats," *Ethology* 117, no. 3 (2011): 217–26, https://doi.org/10.1111/j.1439-0310.2010.01863.x.

28-29 the "default mode network" in our brains: Leo Schilbach, Simon B. Eickhoff, Anna Rotarska-Jagiela, Gereon R. Fink, and Kai Vogeley, "Minds at Rest? Social Cognition as the Default Mode of Cognizing and Its Putative Relationship to the "Default System" of the Brain," *Consciousness and Cognition* 17, no. 2 (2008): 457–67, https://doi.org/10.1016/j.concog.2008.03.013.

30 we pay more attention to high-status people: Nathaniel Ratcliff, Kurt Hugenberg, Edwin Shriver, and Michael Bernstein, "The Allure of Status: High-Status Targets Are Privileged in Face Processing and Memory," *Personality and Social Psychology Bulletin* 37, no. 8 (2011): 1003–15, http://dx.doi.org/10.1177/0146167211407210.

30 activates brain regions involved in feelings of reward: Caroline F. Zink, Yunxia Tong, Qiang Chen, Danielle S. Bassett, Jason L. Stein, and Andreas Meyer-Lindenberg, "Know Your Place: Neural Processing of Social Hierarchy in Humans," *Neuron* 58, no. 2 (2008): 273-283, https://doi.org/10.1016/j.neuron.2008.01.025; PBS News Hour, "Social

Status is Hard-Wired into the Brain, Study Shows," April 25, 2008, https://www.pbs.org/newshour/science/science-jan-june08-status_04-25.

Escaping

31 In their FTC antitrust trial: Meta Platforms, Inc., "Meta's Opening Statement," presentation slides in *FTC v. Meta Platforms, Inc.*, No. 1:20-cv-03590-JEB (April 2025): 21.

31 The majority is spent on videos or Reels: Meta Platforms, Inc., "Meta's Opening Statement," presentation slides in *FTC v. Meta Platforms, Inc.*, No. 1:20-cv-03590-JEB (April 2025): 38; Creators (@creators), "A Couple of Fun Facts: Reels Account for 50% of All Time Spent on Instagram," *Threads*, October 2, 2024, https://www.threads.com/@creators/post/DAor70JxDSH.

32 "drawing away of the mind": Nir Eyal, *Indistractable: How to Control Your Attention and Choose Your Life* (Dallas, TX: BenBella Books, Inc., 2019), 12, Kindle edition.

Isolating Algorithmic Entertainment

40 which is ridiculously hard: Jonathan Lee, "Esports Stars Have Shorter Careers than NFL Players. Here's Why," *The Washington Post*, April 19, 2022, https://www.washingtonpost.com/video-games/esports/2022/04/19/esports-age-retirement/.

Separation

42 The average person checks their phone around 200 times per day: Trevor Wheelright, "Cell Phone Usage Stats 2025: Americans Check Their Phones 205 Times a Day," *Reviews.org*, January 1, 2025, https://www.reviews.org/mobile/cell-phone-addiction/.

42 Often it happens subconsciously as an automatic reflex: Jihwan Park, Jo-Eun Jeong, and Mi Jung Rho, "Predictors of Habitual and Addictive Smartphone Behavior in Problematic Smartphone Use," *Psychiatry Investigation* 18, no. 2 (2021): 118–25, https://doi.org/10.30773/pi.2020.0288.

The Willpower Method

49 studies show that willpower is a finite resource: Roy F. Baumeister, Ellen Bratslavsky, Mark Muraven, and Dianne M. Tice, "Ego Depletion: Is the Active Self a Limited Resource?," *Journal of Personality and Social Psychology* 74, no. 5 (1998): 1252–65, https://doi.org/10.1037/0022-3514.74.5.1252.

50 first popularized by Allen Carr: Allen Carr, *Easy Way to Stop Smoking* (New York: Clarity Marketing, 2011); Allen Carr's Easyway, "About Allen Carr's Easyway," accessed September 14, 2025, https://www.allencarr.com/about-allen-carrs-easyway/.

50 outperforms even gold-standard clinical treatment: Sheila Keogan, Shasha Li, and Luke Clancy, "Allen Carr's Easyway to Stop Smoking – A Randomised Clinical Trial," *Tobacco Control* 28, no. 4 (2019): 414–19, http://dx.doi.org/10.1136/tobaccocontrol-2018-054243.

Pleasure

55 "modern-day hypodermic needles": Anna Lembke, *Dopamine Nation: Finding Balance in the Age of Indulgence* (New York: Dutton, 2021), 1.

55 Dopamine is often blamed, as it's central to addiction: Lembke, *Dopamine Nation*, 49.

55 studies have tied it to internet use: Debasmita De, Mazen El Jamal, Eda Aydemir, and Anika Khera, "Social Media Algorithms and Teen Addiction: Neurophysiological Impact and Ethical Considerations," *Cureus* 17, no. 1 (2025), https://doi.org/10.7759/cureus.77145; Min Liu and Jianghong Luo, "Relationship between Peripheral Blood Dopamine Level and Internet Addiction Disorder in Adolescents: A Pilot Study," *International Journal of Clinical and Experimental Medicine* 8, no. 6 (2015): 9943–48, https://pubmed.ncbi.nlm.nih.gov/26309680/; Aviv Weinstein, Abigail Livny, and Abraham Weizman, "New Developments in Brain Research of Internet and Gaming Disorder," *Neuroscience & Biobehavioral Reviews* 75 (2017): 314–30, https://doi.org/10.1016/j.neubiorev.2017.01.040.

55 *pursuit* of pleasure chemical: Lembke, *Dopamine Nation*, 48-9; Bryon Adinoff, "Neurobiologic Processes in Drug Reward and Addiction," *Harvard Review of Psychiatry* 12, no. 6 (2004): 305–20, https://doi.org/10.1080/10673220490910844.

55 The actual feeling of pleasure may come more from: Stephanie Antons, Matthias Brand, and Marc N. Potenza, "Neurobiology of Cue-Reactivity, Craving, and Inhibitory Control in Non-Substance Addictive Behaviors," *Journal of the Neurological Sciences* 415 (2020): 116952, https://doi.org/10.1016/j.jns.2020.116952; Sergey Yu Tereshchenko, "Neurobiological Risk Factors for Problematic Social Media Use as A Specific Form of Internet Addiction: A Narrative Review," *World Journal of Psychiatry* 13, no. 5 (2023): 160–173, https://doi.org/10.5498/wjp.v13.i5.160.

56 Homeostasis is the body's way of maintaining internal equilibrium: Encyclopaedia Britannica, "Homeostasis," *Britannica*, accessed September 14, 2025, https://www.britannica.com/science/homeostasis.

56 an evolutionary signal—it helps us choose one behavior over another: Kenneth Blum, Marjorie Gondré-Lewis, Bruce Steinberg, Igor Elman, David Baron, Edward J Modestino, Rajendra D Badgaiyan, and Mark S Gold, "Our Evolved Unique Pleasure Circuit Makes Humans Different from Apes: Reconsideration of Data Derived from Animal Studies," *Journal of Systems and Integrative Neuroscience* 4, no. 1 (2018): https://doi.org/10.15761/JSIN.1000191; Peter Sterling, "Allostasis: A Model of Predictive Regulation," *Physiology & Behavior* 106, no. 1 (2012): 5–15, https://doi.org/10.1016/j.physbeh.2011.06.004. The neuroscience term is more precisely "allostasis" for this dynamic, but for broader audience readability I described it through homeostasis.

56 dopamine receptors "downregulate" to compensate: Nora D. Volkow, Gene-Jack Wang, Joanna S. Fowler, Dardo Tomasi, Frank Telang, and Ruben Baler, "Addiction: Decreased Reward Sensitivity and Increased Expectation Sensitivity Conspire to Overwhelm the Brain's Control Circuit," *Bioessays* 32, no. 9 (2010): 748–55, https://doi.org/10.1002/bies.201000042.

56-57 One study expressly measured internet addicts: Sang Kim, Sang-Hyun Baik, Chang Park, Su Jin Kim, Sung Choi, and Sang Kim, "Reduced Striatal Dopamine D2 Receptors in People with Internet

Addiction," *Neuroreport* 22, no. 8 (2011): 407–11, http://dx.doi.org/10.1097/WNR.0b013e328346e16e.

57 This dynamic occurs with the other pleasure chemicals: Laura M. Bohn, Raul R. Gainetdinov, Fang-Tsyr Lin, Robert J. Lefkowitz, and Marg G. Caron, "μ-Opioid Receptor Desensitization by β-Arrestin-2 Determines Morphine Tolerance but Not Dependence," *Nature* 408, no. 6813 (2000): 720–23, https://doi.org/10.1038/35047086; John G. Cottone, "A Better Understanding of SSRI Antidepressants," *Psychology Today*, January 11, 2025, https://www.psychologytoday.com/us/blog/the-cube/202005/a-better-understanding-of-ssri-antidepressants.

57 an established finding in addiction research: Mark Griffiths, "A Components' Model of Addiction within a Biopsychosocial Framework," *Journal of Substance Use* 10, no. 4 (2005): 191–97, https://doi.org/10.1080/14659890500114359; Norman S. Miller, Charles A. Dackis, and Mark S. Gold, "The Relationship of Addiction, Tolerance, and Dependence to Alcohol and Drugs: A Neurochemical Approach," *Journal of Substance Abuse Treatment* 4, no. 3-4 (1987): 197–207, https://doi.org/10.1016/S0740-5472(87)80014-4.

57 receptors can "upregulate" and become more sensitive: Gabriele Scheler, "Regulation of Neuromodulator Receptor Efficacy--Implications for Whole-Neuron and Synaptic Plasticity," *Progress in Neurobiology* 72, no. 6 (2004): 399–415, https://doi.org/10.1016/j.pneurobio.2004.03.008; Christopher E Maggos, Hideo Tukada, Takeharu Kakiuchi, *et al.*, "Sustained Withdrawal Allows Normalization of In Vivo [^{11}C]N-Methylspiperone Dopamine D$_2$ Receptor Binding after Chronic Binge Cocaine: A Positron Emission Tomography Study in Rats," *Neuropsychopharmacology* 19, no. 2 (1998): 146–53, https://doi.org/10.1016/S0893-133X(98)00009-8.

Escape Boredom

63 in that moment all the flowers: Marcel Proust, *Swann's Way* (London: Chatto & Windus, 1922), 62.

64 Research shows that creativity: Guihyun Park, Beng-Chong Lim, and Hui Si Oh, "Why Being Bored Might Not Be a Bad Thing after

All," *Academy of Management Discoveries* 5, no. 1 (2019), https://doi.org/10.5465/amd.2017.0033.

64 One British study gave two groups of people a creativity test: Sandi Mann and Rebekah Cadman, "Does Being Bored Make Us More Creative?," *Creativity Research Journal* 26, no. 2 (2014): 165–73, https://doi.org/10.1080/10400419.2014.901073.

64 numerous examples of thinkers and high achievers: Cal Newport, *Digital Minimalism: Choosing a Focused Life in a Noisy World* (New York: Portfolio/Penguin, 2019), 89–97.

65-66 Jonathan Haidt's *The Anxious Generation* (2024): Jonathan Haidt, *The Anxious Generation: How the Great Rewiring of Childhood Is Causing an Epidemic of Mental Illness* (New York: Penguin Press, 2024).

66 Jean Twenge's *iGen* (2017): Jean M. Twenge, *iGen: Why Today's Super-Connected Kids are Growing Up Less Rebellious, More Tolerant, Less Happy—and Completely Unprepared for Adulthood* (New York: Atria Books, 2017).

66 depression and anxiety more than doubled: American College Health Association, *American College Health Association–National College Health Assessment II: Reference Group Executive Summaries, 2010–2018* (Hanover, MD: American College Health Association, 2010–2018).

66 Suicide rates: Sally C. Curtin and Matthew F. Garnett, "Suicide and Homicide Death Rates Among Youth and Young Adults Aged 10–24: United States, 2001–2021," *NCHS Data Brief*, no. 471 (2023), https://www.cdc.gov/nchs/products/databriefs/db471.htm.

66 According to practically every metric: Zach Rausch and Jonathan Haidt, "The Evidence," *The Anxious Generation* (website), March 2, 2024, https://www.anxiousgeneration.com/research/the-evidence.

Social Value

71 % of people reporting close friends – Pre-Algorithms (2003): Joseph Carroll, "Americans Satisfied With Number of Friends, Closeness of Friendships," *Gallup News Service,* March 5, 2004, https://news.gallup.com/poll/10891/americans-satisfied-number-friends-closeness-friendships.aspx.

71 % of people reporting close friends – Post-Algorithms (2023): Isabel Goddard, "What Does Friendship Look Like in America?," *Pew Research Center*, October 12, 2023, https://www.pewresearch.org/short-reads/2023/10/12/what-does-friendship-look-like-in-america/.

71 Daily time spent with friends: Youyou Zhou, "Want More Time with Friends? It's the Ordinary Things That Matter," *The Washington Post*, December 23, 2024, https://www.washingtonpost.com/opinions/interactive/2024/friends-loneliness-solitude-friendships/. Visualizations of data from the American Time Use Survey.

71 Meta's courtroom presentation: Meta Platforms, Inc., "Meta's Opening Statement," presentation slides in *FTC v. Meta Platforms, Inc.*, No. 1:20-cv-03590-JEB (April 2025): 21.

73 People who are popular in real life: Carlos Jennewein, Annika Baumann, and Stefan Lessmann, "To Use or Not to Use: The Relationship between Personality Traits and Instagram Usage," paper presented at the *Wirtschaftsinformatik 2020 Conference*, Potsdam, Germany, March 8–11, 2020, https://doi.org/10.30844/wi_2020_o3-jennewein; Thomas Bowden-Green, Joanne Hinds, and Adam N. Joinson, "How Is Extraversion Related to Social Media Use? A Literature Review," *Personality and Individual Differences* 164 (2020): 110040, https://doi.org/10.1016/j.paid.2020.110040.

74 brain is bad at comprehending large numbers: Yasemin Saplakoglu, "Why the Human Brain Perceives Small Numbers Better," *Quanta Magazine*, November 9, 2023, https://www.quantamagazine.org/why-the-human-brain-perceives-small-numbers-better-20231109/.

74 evolved over millions of years to be incredibly adept: Esther Herrmann, Josep Call, María Victoria Hernández-Lloreda, Brian Hare, and Michael Tomasello, "Humans Have Evolved Specialized Skills of Social Cognition: The Cultural Intelligence Hypothesis," *Science* 317, no. 5843 (September 7, 2007): 1360–66, https://doi.org/10.1126/science.1146282.

Knowledge

79 An irresistible urge to learn: Michael Balter, "Did Big Brains Sap Our Strength?," *Science*, May 7, 2014, https://www.science.org/content/article/did-big-brains-sap-our-strength; Celeste Kidd and Benjamin Y. Hayden, "The Psychology and Neuroscience of Curiosity," *Neuron* 88, no. 3 (2015): 449–60, https://doi.org/10.1016/j.neuron.2015.09.010.

79-80 Over 20 million videos uploaded to YouTube: YouTube, *YouTube for Press*, accessed September 16, 2025, https://blog.youtube/press/.

80 500 million posts on X: Sarah Perez, "Actually, X Sees 500M Posts Per Day — Not 100M-200M as Musk Recently Said," *TechCrunch*, October 4, 2023, https://techcrunch.com/2023/10/04/actually-x-sees-500m-posts-per-day-not-100m-200m-as-musk-recently-said/.

80 estimated 270 million posts on TikTok: Benjamin Steel, Miriam Schirmer, Derek Ruths, and Juergen Pfeffer, "Just Another Hour on TikTok: Reverse-engineering Unique Identifiers to Obtain a Complete Slice of TikTok," arXiv, May 13, 2025, arXiv:2504.13279, https://doi.org/10.48550/arXiv.2504.13279.

80 The "Flynn Effect": Denise Andrzejewski, Elisabeth L. Zeilinger, and Jakob Pietschnig, "Is There a Flynn Effect for Attention? Cross-Temporal Meta-analytical Evidence for Better Test Performance (1990-2021)," *Personality and Individual Differences* 216 (2024): 112417, https://doi.org/10.1016/j.paid.2023.112417; Robert J. Sternberg and James C. Kaufman, "Intelligence," in *Encyclopedia of the Human Brain*, ed. V. S. Ramachandran (San Diego: Academic Press, 2002), 587–97, https://doi.org/10.1016/B0-12-227210-2/00179-5.

81 has been *reversing*: James R. Flynn and Michael Shayer, "IQ Decline and Piaget: Does the Rot Start at the Top?" *Intelligence* 66 (2018): 112–21, https://doi.org/10.1016/j.intell.2017.11.010; Elizabeth M. Dworak, William Revelle, and David M. Condon, "Looking for Flynn Effects in a Recent Online U.S. Adult Sample: Examining Shifts Within the SAPA Project," *Intelligence* 98 (2023): 101734, https://doi.org/10.1016/j.intell.2023.101734.

81 one Norwegian study: Bernt Bratsberg and Ole Rogeberg, "Flynn Effect and Its Reversal Are Both Environmentally Caused," *Proceedings of the National Academy of Sciences* 115, no. 26 (2018): 6674–78, https://doi.org/10.1073/pnas.1718793115.

81 From 2003 to 2022, the average scores recorded by PISA: OECD, *PISA 2022 Results (Volume I): The State of Learning and Equity in Education* (Paris: OECD Publishing, 2023), https://doi.org/10.1787/53f23881-en.

82 the American NAEP assessments: U.S. Department of Education, Institute of Education Sciences, National Center for Education Statistics, *NAEP Long-Term Trend Assessment Results: Reading and Mathematics for Age 13*, The Nation's Report Card, 2023, https://www.nationsreportcard.gov/ltt/?age=13.

83 Monitoring the Future survey: John Burn-Murdoch, "Have Humans Passed Peak Brain Power?," *Financial Times,* March 14, 2025, https://www.ft.com/content/a8016c64-63b7-458b-a371-e0e1c54a13fc; University of Michigan, Institute for Social Research, *Monitoring the Future: Public-Use Cross-Sectional Datasets* (Ann Arbor, MI: Interuniversity Consortium for Political and Social Research), accessed September 8, 2025, https://www.icpsr.umich.edu/web/NAHDAP/series/35.

84 "if the book is completely intelligible to you": Mortimer J. Adler and Charles Van Doren, *How to Read a Book* (New York: Touchstone, 1972), 7, Kindle edition.

85 "working memory is the mind's scratch pad": Nicholas Carr, *The Shallows: What the Internet Is Doing to Our Brains* (New York: W. W. Norton & Company, 2020), 123–25.

85 one 1956 study suggested: George A. Miller, "The Magical Number Seven, Plus or Minus Two: Some Limits on Our Capacity for Processing Information," *Psychological Review* 63, no. 2 (1956): 81–97, https://doi.org/10.1037/h0043158.

85 the "cognitive load": John Sweller, *Instructional Design in Technical Areas* (Camberwell, Australia: Australian Council for Educational Research, 1999), 4–5.

85 "unable to retain the information": Carr, *The Shallows*, 125.

87 "How often does it occur": Postman, *Amusing Ourselves to Death*, 68.

The Apathy Method

95 whether in your conscious or subconscious mind: Nicholas Carr, "How Smartphones Hijack Our Minds," *The Wall Street Journal*, October 6, 2017, https://www.wsj.com/articles/how-smartphones-hijack-our-minds-1507307811.

Beyond Brain Rot

98 As Earl Nightingale once said: Earl Nightingale, "The Strangest Secret," YouTube video, posted by Miracle Malini, January 21, 2023, https://www.youtube.com/watch?v=l1gXZu1i8TM.

100 "Medicine, law, business, engineering, these are noble pursuits": *Dead Poets Society*, directed by Peter Weir (Burbank, CA: Buena Vista Pictures Distribution, 1989).

Acknowledgements

I drew on many sources in writing this book, but I am particularly grateful for the works of Cal Newport—*Digital Minimalism* had a profound effect on me when I first read it in college and sparked my interest in the topic. I am also thankful for the books, seminars, and videos of the late Allen Carr. He has helped millions of people through addiction and inspired my own escape method.

Of course, my deepest thanks go to my parents, Johannes and DD, and my brother, Hans—the most wonderful family I could have dreamed of—for their support and advice at every stage of the writing process. I am also grateful to my cousin Isabelle Stillman, whose editing skills greatly improved the book. And finally, a big thank you to all my friends who generously read early drafts and provided feedback—too many to name here, but you know who you are. The battle against brain rot is, after all, a team effort!